Autism Friendly

Designed as a guidebook for leaders at the beginning of their journey embracing neuroinclusion, *Autism Friendly Business: Serving Neurodiverse Customers* will provide business owners, executives, managers, team members, and associates the tools to integrate strategies and techniques that will enhance their business, while improving the delivery of a quality experience for all. Autistic individuals often experience barriers when engaging with businesses. This book provides solutions and examples on how leaders can remove obstacles to develop supportive and inclusive environments.

Autism Friendly

Jennifer Percival

All people should have access to what is available in their community and beyond. Neurodiverse individuals often experience barriers when engaging with businesses, even when obstacles can be easily remedied. This book . will provide business owners, leaders, managers, team members, and associates the tools to integrate strategies and techniques that will enhance and improve the delivery of a quality and inclusive experience to neurodiverse customers.

Autism Friendly Business
Jennifer Percival 2024

Autism Friendly Business
Serving Neurodiverse Customers

Dr. Jennifer Percival

Routledge
Taylor & Francis Group
A PRODUCTIVITY PRESS BOOK

Cover image Credit: Andrew Blitman

First published 2024
by Routledge
605 Third Avenue, New York, NY 10158

and by Routledge
4 Park Square, Milton Park, Abingdon, Oxon, OX14 4RN

Routledge is an imprint of the Taylor & Francis Group, an informa business

ISBN: 978-1-032-02832-3 (hbk)
ISBN: 978-1-032-02831-6 (pbk)
ISBN: 978-1-003-18540-6 (ebk)

DOI: 10.4324/9781003185406

Typeset in Garamond
by MPS Limited, Dehradun

For Jeremy and EmJ

Cover Illustrator

The cover was illustrated by Andrew Blitman, an artist with autism, who resides in South Florida. Andrew earned a Master's Degree in Marine Conservation Science & Policy from the University of Miami. He served on the UM-NSU Center for Autism and Related Disabilities constituency board from 2011 to 2017. Andrew currently works as a painter, tutor, and mentor, and has served on the Els Center for Autism advisory board since 2018. He is the author of a memoir about his life, and several works of poetry. You can find out more about Andrew by visiting his website at andrewblitman.com

Contents

PART 2 THE WHO: PEOPLE WITH AUTISM, LEADERS, AND COMMUNITY PARTNERS

PART 3 THE WHAT: CREATING CHANGE

Acknowledgments

My career is dedicated to the practice of tikkun olam, healing the world by contributing my experience and energy to our communities, with the goal of making them more inclusive, welcoming, and diverse. This book is part of my desire to establish a long-lasting sustainable societal impact, and it is part of a series of books that originate from the creation and implementation of the Autism Friendly movement.

The Autism Friendly movement was founded and launched from a belief that with just a little education and a few strategies, organizations have the power to make a significant difference and increase the quality of life for many. This initiative would not be possible without the originating visionaries, who experienced and observed barriers to access and inclusion in their communities. These individuals were interested in creating change to shift those exclusionary practices. While this book is about Autism Friendly Businesses, the core of the movement began with an Autism Friendly City. In the Southeast region of Florida, at the City of Boynton Beach, the ADA (Americans with Disabilities Act) Coordinator who first took on this challenge was Debbie Majors. Using the available resources, and crafting some that didn't yet exist, she gathered a team together, and began the journey of enhancing her community. I commend and acknowledge her conviction, resolution, and determination to make her city genuinely inclusive and Autism Friendly.

Debbie Majors was the leader of this movement, and to help her achieve her vision, she developed and utilized community partnerships. I had the honor of being a member of her team, and am proud to have contributed to the process. This dedicated and talented group included Stephanie Soplop, the current ADA Coordinator, Ted Goodenough, the ADA Coordinator Assistant, both at the City of Boynton Beach, as well as the entire ADA Advisory and Inclusion Team which consisted of representatives from local human services and disability-related agencies and organizations. The city leadership, including the Mayor and Commissioners, sustained and supported this effort. The City of Boynton Beach serves as a model Autism Friendly City, thanks to Debbie Majors and her team that worked diligently to improve the quality of life for neurodiverse individuals and their families. This model can be applied to other cities, as well as businesses, organizations, and non-profits.

Florida Atlantic University's Center for Autism and Related Disabilities (FAU CARD) is central to the Autism Friendly movement. The Executive Director, Dr. Jack Scott, and the Director, Dr. Maryellen Quinn-Lunny, have been instrumental in guiding and advocating for this program, and they encouraged my ongoing work in this arena. The FAU CARD team of Clinical Specialists provide training and consultation to our local Autism Friendly organizations, and their expertise and contribution to this work is highly valued. University of South Florida's CARD Center was influential in developing an Autism Friendly Business program, which the FAU CARD program was initially inspired by. Their efforts created the foundation needed to support the growth of the Autism Friendly movement to a national stage.

Part of my legacy is to do everything in my power to leave this world a better place for the next generation. Jeremy and

EmJ, my children, you fill my life with light and inspiration. Thank you for being a part of my purpose and for supporting my dreams and goals. My parents, Ellen and Jerry, and Don and Susan, across my lifetime you've cheered me on and believed my abilities would contribute and make a difference. Thank you for your continuous encouragement. Brian, my incredible brother, I appreciate your steadiness and ongoing support as I've been on my journey. To my lifelong friend, Erica, thank you for being my ride-or-die for over 40 years! Emily, you see me and love me with a fierce steadiness for which I am daily grateful. Impossible things are happening every day! For all of the love in my life, that lifts me up and reminds me why my words matter, I thank you.

About the Author

Dr. Jennifer Percival is the author of *Autism Friendly Cities: How to Develop an Inclusive Community* and *Families Creating Employment Opportunities for Individuals with Developmental Disabilities: Understanding the Contribution of Familial Entrepreneurship*. She is the Associate Director of Florida Atlantic University's Center for Autism and Related Disabilities. She holds a Doctorate from FAU in Educational Leadership and Research Methodology with a research area of Disability Studies. She is a Nationally Certified Rehabilitation Counselor, DEI expert, advisor on disability and employment strategies, and has extensive experience working with transition-aged students and their families. Dr. Percival is a sought-after speaker, educator, and consultant.

Dr. Percival is the founder of the Autism Friendly Cities and Autism Friendly Business programs at FAU CARD. She is the creator of the world's only Autism Friendly Cities Institute. Dr. Percival is an award-winning researcher in disability studies, and she participates on national and regional boards, work groups, and committees that guide cities and business through the Autism Friendly process. Jennifer serves on the Royal Caribbean International Autism Advisory Board, the CareerSource Palm Beach County Youth Employment Board, the City of Boynton Beach Inclusion Team, and Palm Beach County's Special Needs Advisory Council. Dr. Percival was the

2019 Florida Graduate Research Symposium second-place state-wide winner for her dissertation research. She was the State of Florida's Workforce System's Barbara Griffin Workforce Excellence Award winner in 2017, FAU's Lifelong Learning Society Lunafest Scholarship recipient in 2017, and FAU's College of Education North Campus Student of the Year in 2016.

Prior to working at FAU, Dr. Percival divided her career between the Florida Department of Education's Division of Vocational Rehabilitation, and the Florida Department of Children and Families. Dr. Percival is a community leader who develops infrastructure through her work with collaboration and regional planning, while assisting professionals and individuals with program improvement and strategic implementation.

*Disclosure statement: The content of this book is the opinion of the author and is not representative of FAU CARD, the FLDOE, or any other entity.

Introduction

Businesses are now engaging Diversity, Equity, and Inclusion (DEI) Specialists who develop and implement strategies that grow and champion diverse environments. DEI practice has the power to create meaningful change that is inclusive of expanding ethnic and cultural diversity, examining equity, exploring gender roles, social justice actions, and widening disability inclusion. A focus on DEI shifts, strengthens, and amplifies the dimensions of the internal lens of an organization, which can be leveraged into increased financial performance and expanded reach to customers. As such, your own DEI practice must address neurodiversity and autism.

The prevalence of autism is so significant that it cannot be ignored by businesses. In the United States, 1 in 44 individuals is diagnosed with an autism spectrum disorder (Center for Disease Control, 2021), and the World Health Organization (2022) indicates that 1 in 100 children globally has autism.

Neurodiverse customers are significant contributors to the economy. They are children, teenagers, young adults, older adults, and everything in between, who will vary in cultural background, religion, ethnicity, location, and socioeconomic level. There are also those that exist in the sphere of neurodiverse people, including parents, grandparents, friends, colleagues, and neighbors, who may be caregivers or supports. All of these individuals are consumers or consumer-support

DOI: 10.4324/9781003185406-1

people that are your potential customers. As a business leader, it is necessary to have a strong understanding of what their needs will be as they interact with your company.

As an author, my job is to enroll you in my words and ideas. It is a shame that inclusive practice is not an automatic aspect of development and design, and that it continues to require some level of recruiting others that the time and resources are valuable financially, socially, but societal growth does take time. I am excited to see that you and others are engaging in these conversations and taking action at organizational levels.

Businesses come in all types of forms, which can range from a grocery store to a bank to a restaurant. Every company that offers a product or service will be frequented at some point by an autistic person and their families. For ease of identification throughout this book, we will call them neurodiverse or autistic consumers and customers, and it is those consumers who are neurodiverse, that we must consider when developing the experience that they will have when interacting with a business.

As you read on, please frame this text in the concept that your decisions, as a business owner, leader, or even the person stocking the shelves or running the cash register, have the power to impact lives. The way you prepare for your customers and how you respond to them may even dictate the future of your business and contribute to your level of success. Aspects of day-to-day operations that you consider minor, from paint colors and directional signs, to the use of visual supports, and website design are all areas to consider. Training and educating your team, from front-line employees, through the C Suite, will be essential. By taking these first steps, you are being a proactive leader who serves as a model for other businesses, raising the tide that lifts all boats and works towards creating an inclusive, accessible, and neurodiverse community.

Please feel free to contact me to ask questions or for consultation at drjenniferpercival@gmail.com

WHY? 1

While writing this book, I was sitting on a plane next to a family that was coming back from New York Comic Con. The Mom had exceptionally cool black high tops adorned with Mickey Mouse head crystals, and her adult son, probably in his early 20s, was carrying a bag of Marvel swag and goodies. The parents were talking across the aisle about how the trip went and the ups and downs they experienced. They were thankful they had their son's headphones with them as the convention center was packed and loud, they said they needed a hotel room closer to the center next year so they could take more breaks, and they were super happy about the way-finding techniques used by the convention planners on the app which kept their son in charge of where they were headed next and allowed him to manage the appearance schedule. I was smiling as I listened to them and thought about other ideas that the convention planners might have implemented to make the con a bit more accessible to neurodiverse attendees. As I was musing on this, the flight attendant asked our row what we would like to drink. When she got to the young man, who was wearing noise-blocking ear covers on the plane, she provided the list of drink options verbally, and then showed him the pictures of the drinks on her tablet. He made his choice independently by pointing to the soda he wanted, which he was promptly served.

DOI: 10.4324/9781003185406-2

When we explore the possibilities of bringing a more inclusive and open experience to neurodiverse customers, there are times businesses feel overwhelmed with what they believe they need to put in place. However, these few examples demonstrate that it is the small changes, and the flexibility in communication, that can open up opportunities for those that would otherwise experience a barrier. This is our 'Why,' our 'Why we are here.' We take on this work to broaden and improve the quality of life for everyone that engages with our businesses. It is important and necessary because we become companies that support equal access and we develop relationships with our customers that are inclusive of meeting their diverse needs.

Chapter 1

Why Us? and Where Are We Now?

Some businesses that are just beginning to look at neuroinclusion ask 'Why us?' wondering why neurodiversity is their responsibility. The simplest answer is, because it will make both your business and your community better. If not us, then who? It takes a gigantic effort to create societal change, but that weight is more easily carried when it is distributed amongst many. These many must also be in different geographical regions, serve various types of communities, and represent differing business sectors. We will not push our culture forward if only one type of business in one city takes this on. It must be a larger effort and that is why we, as a global community, are asking for your help and participation.

As you explore your business' initial dive into being Autism Friendly, there are internal conversations to be had. These are ways to spark ideas and to begin the process of self-evaluation. Knowing where you are in the process is key. There may be strategies that are inherently inclusive which you already practice. You may also uncover barriers customers experience that you did not know existed. Knowledge is power, and

DOI: 10.4324/9781003185406-3

recognizing where you are on the path is the first step to being on the journey and deciding where you are headed.

*Reflect and Discuss

Consider These Questions as Conversation Starters:

- Share a positive experience about a time a person with autism had a successful interaction with your business.
- Discuss a positive personal experience of working directly with a person with autism in your capacity at your organization.
- If there are autism-specific or neurodiverse components to the customer experience, are they inclusive, or do they segregate people with autism or a disability to groups where they are only with others who also have autism? (for example, a bookstore may offer sensory friendly story hour)
- Has a person with autism reported an experience where they felt excluded or experienced barriers to services you offer?
- Do you have employees that have requested to limit participation or access at your business attendance to people without autism or neurodiversity?
- Have there been complaints by adults, parents, or families about limited inclusive opportunities?
- Are all of the experiences you offer open to anyone with any level of need or support? If not, what are your limitations, and why?
- Has there been a negative incident related to a person with autism or a disability interacting with your organization? If so, what steps were taken after the incident to remedy the situation? Did those steps resolve the issues, or were they a temporary fix? If not, what type of incident are you concerned may happen?

- Has there been a poorly handled situation at your company related to a person with autism that resulted in media coverage? If so, how did your business respond to that negative presentation to the community? What was done to prevent it from happening again?

DEIA: Diversity, Equity, Inclusion, and Accessibility

As we discuss our experiences and begin to develop lenses that promote inclusion, we must discuss DEIA: Diversity, Equity, Inclusion, and Accessibility. The A, Accessibility, was recently added to this philosophy. In 2021, via Executive Order 13985, the United States Federal Government issued the following definitions for these terms:

- The term 'diversity' means the practice of including the many communities, identities, races, ethnicities, backgrounds, abilities, cultures, and beliefs of the American people, including underserved communities.
- The term 'equity' means the consistent and systematic fair, just, and impartial treatment of all individuals, including individuals who belong to underserved communities that have been denied such treatment.
- The term 'inclusion' means the recognition, appreciation, and use of the talents and skills of employees of all backgrounds.
- The term 'accessibility' means the design, construction, development, and maintenance of facilities, information and communication technology, programs, and services so that all people, including people with disabilities, can fully and independently use them. Accessibility includes the provision of accommodations and modifications to ensure

equal access to employment and participation in activities for people with disabilities, the reduction or elimination of physical and attitudinal barriers to equitable opportunities, a commitment to ensuring that people with disabilities can independently access every outward-facing and internal activity or electronic space, and the pursuit of best practices such as universal design.

The practice of DEIA philosophy is being integrated into many organizations, workplaces, and mission statements. This incorporation of DEIA can be discussed as an aspect of strategic planning, training curriculum, ways of work, and as a lens of doing business internally and when public facing. An organization called Disability, which offers training and a network to businesses that support inclusive business practices, talks about the DEI Index as an instrument, and explains the process: 'The Disability Equality Index (DEI) is a comprehensive benchmarking tool that helps companies build a roadmap of measurable, tangible actions that they can take to achieve disability inclusion and equality. Each company receives a score, on a scale of zero (0) to 100, with those earning 80 and above recognized as "Best Places to Work for Disability Inclusion."' A key to DEIA is to embrace an uncompromising level of self-assessment, personally and globally, as well as creating safe spaces for difficult conversations and meaningful problem solving, where every voice is heard. Thoughtful conversations are needed when asking yourself, your leaders, and your customers about times they feel excluded, or marginalized.

*Evaluate Your Business

Utilizing either a specific department, or a more global approach to your business at large, kick the tires and have a look under the hood. To begin, discuss what you know, consider what you don't know, and begin to have difficult conversations. The questions below will help you initiate a dialogue.

- Do you know how many consumers living in the region(s) served by your business that are neurodiverse or on the autism spectrum?
- What strategies or accommodations do you currently offer to people with disabilities?
- Of these accommodations, are any of them designed to serve neurodiverse or autistic consumers?
- What internal policies or initiatives are currently in place to support the inclusion of autistic consumers at your business?
- If no policies are in place, what is your most pressing concern related to serving neurodiverse customers?
- Has your leadership or public-facing employees ever received any training about people with disabilities and what their needs might be? If so, did this training include information about neurodiverse consumers?
- Are you aware of organizations and agencies in your region that support people with autism? If so, what are they and how might they help your business be more inclusive?
- How many of your own employees are neurodiverse or autistic? Do any of these employees participate in work groups, ERGs, task forces, or contribute to the design of the business or how customers are served?
- How many of your customers are neurodiverse? (Some businesses may have these statistics based on what they offer their customers, i.e. a cruise line may have guests self-disclose to request specific supports or accommodations.)
- Do the community boards or professional associations you participate in or belong to have neurodiversity in representation?

- What worries you the most about not being an Autism Friendly Business?
- Have you experienced a situation where a person with autism was excluded or was unable to utilize your services or products due to needed supports not being in place?
- If you could make one change immediately, to increase inclusion, what would that change be, and why?

Chapter 2

Why We Need to Engage Others

You may be an executive-level decision-maker at your business, a middle manager, or a front-line worker. Regardless of your position, there are many opportunities to implement Autism Friendly strategies, and across all levels, there will be a need to engage others. Most designated DEI managers, ADA coordinators, and HR specialists are typically the professionals who decide to pick up this book with a vision of a more neurodiverse business and customer base, but it is important to help all team members embrace new initiatives to support growth.

In order to reach everyone, we must encourage those that are not connected to the neurodiverse community. That enrollment comes with being aware of and knowing and serving people with autism. Engagement occurs through education, sharing our experiences, and highlighting our success. We need to relay what we try and implement, and what works, and what doesn't. It is okay to communicate failures because that may save another group from making mistakes. The effort and attempt is what is important in those moments.

The people you may need to enroll will vary. If you are a staff person at a business, you may need to share information

DOI: 10.4324/9781003185406-4

with your leadership team. Leaders may need to convince high-level executives that the resources spent will be worthwhile. If you are a business owner, you may even need to help your current customers and patrons understand why you are making changes and how they will promote a more inclusive experience. There will be roadblocks in this process because there will be people who don't immediately understand how far the impact and benefits are.

What Is the Time Commitment?

The time requirements to become Autism Friendly are dependent on the level of commitment of your business is ready to offer. If your availability is minimal, that is okay because there are many ways to begin, and some do not need excessive time or effort. One way to begin is to pick a single department or program, such as an ERG or an HR department, and work through the steps to make it Autism Friendly. Another opportunity is to explore programs at your business that may already work with children, for example, your company supports or sponsors a high school robotics team or little league. Training to work with children may already be required, and adding information about neurodiversity could be easily introduced.

How Do We Pay for This?

Many of the strategies discussed in this book come at little to no cost. For example, to create a visual support can be as simple as using your computer software and printing on paper to post the visual. Slightly added expenses might be taking that visual and hanging it in an inexpensive picture frame or a plastic stand. Other ways of keeping costs down are to use items that you already own and that are part of the space. For example, if you have a television that is in your entrance area or lobby, adding information and visual supports that are Autism Friendly to that television might be a no-cost solution.

There are times when additional resources are required, which is most often to cover the hours needed for training and implementation. For some, it may be easy to incorporate inclusive strategies as part of work responsibilities or tasks, for others the focus is on increasing knowledge through education. An easy way to start this process is to have training included in the hiring and onboarding process for new employees. There may be times where a task is slightly outside the scope of your employees and possibly covering salaries during these times may have a small cost.

Larger cost items might include purchasing furniture, redesigning a physical space, or adding items like tents used as a quiet space, to an event. Other costs could include things on a large scale like redesigning a park, or a natural area. However, when these remodels are being done as part of normal maintenance, it is easier to incorporate strategies at that time and receive input from experts as part of the planning process. For example, if you are experiencing a redesign or new construction, it is useful to bring in an ADA or inclusion team during the planning process so that it is more cost-friendly to incorporate strategies and spaces upfront, versus remodeling afterwards.

There are some ways practical and innovative ways to cover costs for this initiative. Some organizations reach out to different funding sources and apply for grants. There are many national and international programs that offer funds in the form of neighborhood or community project supports. Another way that funding can be located is to look at your budget and evaluate if any funds are able to be reallocated. For example, at the City of Boynton Beach, they decided to utilize some of the funds from the parking tickets that are issued to people who illegally park in disabled parking spaces. This was an ingenious idea, because it redirected money collected when rights were violated and access blocked, and brought it back to the very community those spaces were intended to serve.

Some businesses have already demonstrated their commitment to neurodiversity by hiring an ADA or DEI specialist, or an inclusion coordinator. Sometimes these are full-time positions that are responsible for an entire organization, or they are positions that are incorporated into other roles, perhaps for a certain number of specified hours per week. Having this type of person on staff with job responsibilities that include maintaining updated laws and policies, and facilitating the Inclusion Team, as well as working closely with customers, can be an excellent opportunity for your business to show its commitment and to have a dedicated person to work through these processes.

Financial Benefits

When exploring implementing neurodiverse and Autism Friendly strategies, it is okay to consider monetary advantages. There are benefits to being both socially responsible and inclusive. When a business is open to all customers, and takes the time and resources to remove barriers and improve the experience, it creates a rapport and develops loyalty. When neurodiverse customers feel comfortable and have safe interactions with your company, they are more likely to utilize your products and services, and to recommend them to others. Being Autism Friendly expands reach and broadens the customer base.

When You Meet with Resistance

In my experience working with businesses, there are some leaders that express interest in being Autism Friendly, but that are not fully ready to engage with the entire process. This tends to be the case especially with larger organizations that feel overwhelmed at training so many staff and developing the strategies required to go along with the implementation. One way to help navigate this is to select one group or department to engage to begin, typically a smaller number of employees. They then follow the same process of training, interacting with community agencies for consultation, and then implementing a few strategies for improved access, but on a much smaller scale. For example, this might look like creating an Autism Friendly swimming program at the gym you manage. The swim teachers and coaches, along with the lifeguards at the pool, are trained about autism, visual supports are created for the physical space, and a social story is developed for children that are new to using the pool or that are first-time swimmers. The consultation for the department may be a quarterly meeting with a local autism expert who provides suggestions on working one-on-one with children who might exhibit behaviors. In this case, it would not be that the entire gym is designated as Autism Friendly, but that there is an Autism Friendly swimming program within the business. What frequently occurs is that once there is a department that has completed the process, other groups will have a better feel about the commitment of resources, and the impact, which can be motivating. There are many businesses that choose this path and over time become Autism Friendly, by completing the process one department at a time.

Chapter 3

Why Listening Is Powerful

Being an inclusive business demonstrates that you are committed to gaining a greater understanding of those that you serve. Part of that expression is listening and acquiring information about and from your customers. All people want to feel heard and they are eager for the opportunity to contribute their thoughts, ideas, and concerns. One method of listening to your customers' experiences and perspectives is surveying them. You can develop questionnaires that may be used prior to engagement with a customer, as a tool to gain knowledge or prepare for their needs, or after an interaction as examining satisfaction.

Customer Information Questionnaire

A questionnaire can be used to help guide and shape the customer experience prior to an individual engaging your service or product. For example, in hospitality or transportation, such as a hotel, airline, or cruise line, asking questions up front can help reduce barriers and prevent uncomfortable situations.

DOI: 10.4324/9781003185406-5

Questionnaires can be very brief and generalized, or more specific. Here are some sample questions to consider.

- Do you (or your child) require any accommodations? If yes, please indicate how we can best support your needs.
  - Some organizations provide a list of options on their form.
- Will you be bringing any equipment with you? If yes, is there anything we can provide that will support use of the equipment?
  - Businesses may be able to offer charging stations, extension cords, filtered water, hypoallergenic bedding, etc.
- What are your safety concerns, and how can we support a safe environment for you or your loved one?
  - Offering door alarms, bed rails, life jackets, extra pillows, or other items may be useful.
- Do you or your child have any specific dietary requests?
  - This may include allergies, gluten-free options, specific types or brands of food, lactose-free milk, specific utensils.
  - Some autistic individuals have a list of safe foods that they are comfortable with in most settings, and that they depend on. If those foods are not offered by your business, the family may need notice and support to bring those foods with them.
- Are you or your child sensitive to noise or light?
  - This may guide room or seat selections or recommendations, as well as overhead or lamp lighting.
  - This is also an opportunity to encourage customers to bring headphones or earplugs with them if your business has high volume or stimulating experiences.
- How can we best communicate with you?
  - This may be a request for email or text only, phone, interacting with a guardian or loved one, or utilizing

visual supports. Also consider types of quiet or low sensory spaces you can offer.

- Please share more about how we can support you in having a positive experience interacting with our business.
  - This question is open-ended and may not always be an option to include, but if you are at the beginning of implementing supports, it is recommended to offer space for customers to provide more information that you can use to shape your program development.

Customer Satisfaction Survey

*A customer satisfaction survey can help you gather baseline data about a person's experience at your business as a customer. Responses can guide organizational change, or help shape customer experience questions needed to improve accessibility.

- How long have frequented this business?
- How often do you utilize these services?
- What type of products/services do you utilize?
- Do you find the directions related to products/services useful and easy to understand?
- What programs and activities do you participate in that are offered by our business?
- Does our business make you feel welcomed?
- Are there accommodations that you've requested which we provided for you?
- Have you requested accommodations that we were not able to provide?
- Are you able to successfully and easily navigate our website? If not, how can we improve this for you?

- When you communicate with staff, are you able to easily understand the information they provide? If not, please suggest how we can do this better.
- Do you believe our staff have a good understanding of people with autism?
- Have you had an experience where you felt like you could not access our products, services, or programs? If yes, please provide a description for us.
- Do you find our organization's leadership approachable?
- Would you be willing to participate in an inclusion team that advises our business on best practice, or are you willing to provide us with feedback in a one-on-one environment?

2

THE WHO: PEOPLE WITH AUTISM, LEADERS, AND COMMUNITY PARTNERS

It is difficult to develop strategies and supports without knowledge, and it is impossible to have a successful initiative in isolation. We cannot create change without information, and we definitely cannot do it alone. This section will provide a brief overview of autism, it will discuss training opportunities and how to determine what your needs are, and it will teach you how to connect with trainers in your community. This part will also guide you through the process of putting together an inclusion team and how to connect with community organizations. Once that team is in place, it will guide you on how to utilize the experience and expertise of those team members and how to reap the benefit of those collaborations.

DOI: 10.4324/9781003185406-6

Chapter 4

Let's Talk a Bit about Autism

I am different, not less. The world needs all types of minds.

– Temple Grandin

This chapter will contain a general overview of autism. People say that when you've met one person with autism you've met one person with autism. This is absolutely true. While there are some overarching commonalities and similar issues that this group face, every person with autism will have their own abilities, support needs, personality, and goals. To help you assess your team's learning, a short quiz that can be utilized as a pre-test and/or post-test is provided. In this chapter, we will review these areas and others; basic diagnostic information about autism, the rates and prevalence of autism, the characteristics that some people with autism may present with, difference versus deficit, and neurodiversity. My goal with this chapter is to bridge the understanding of the diagnosis to those that are

DOI: 10.4324/9781003185406-7

inexperienced or unfamiliar with the needs and general characteristics of individuals with autism. As such, this chapter will not contain highly academic or medical data, but the foundational information for you as you develop an Autism Friendly and neurodiverse initiative.

*Quiz (Pre/Post Test): A Brief Quiz on Autism

1. What is the prevalence of autism according to the U.S. CDC?
   a. 1 in 1,730
   b. 1 in 260
   c. 1 in 44
   d. 1 in 98
2. Is autism an intellectual disability?
   a. Yes
   b. No
3. There is a blood test to diagnose autism.
   a. True
   b. False
4. Autism is more prevalent in boys.
   a. True
   b. False
5. What are the three main areas of difference for individuals with autism?
   a. ______________
   b. ______________
   c. ______________
6. Neurodiversity includes only autistic people.
   a. True
   b. False
7. Being different means that you have a deficit.
   a. True
   b. False

Basic Diagnostic Information

When we talk about autism spectrum of disorders, it is important to understand that autism is a range, a scale, a continuum. That each person who has a diagnosis of autism will experience different aspects and characteristics of autism. That an individual may have some facets of autism, but not others, and may experience a varying intensity of specific parts of Autism. It will present in a unique way for everyone.

Autism is a developmental disability, and not an intellectual disability. This is an important distinction, because there are some people that think autism is related to intelligence. This is not the case. There may be individuals that have both autism and an intellectual disability, but they are different diagnoses. People with autism may also have co-occurring diagnoses where they experience autism and a mental illness or physical issue. It is possible that sometimes that co-occurrence presents more significantly than the autism, or vice versa. For example, a person with autism might also have a diagnosis of depression. There may be times during that person's life where their autism is more pronounced, and other time where their depression is more prominent.

Autism can be diagnosed at any time in life, but it is most frequent to see the diagnosis come during early childhood. A diagnosis often results from a family's interaction with a team of practitioners, which may be in consultation with a child's pediatrician, specific service provider such as a speech and language pathologist, or through an early intervention or school-based team. When considering the diagnostic criteria of autism is important to recognize that is a complex developmental disorder, that has no known cause, no current medical test for diagnosis (such as a blood test), there are no ethnic racial, or cultural boundaries, and no socioeconomic boundaries. Early diagnosis and intervention, when possible, is best practice and most supportive of the child's needs throughout early development and education.

Prevalence

At the time of the writing of this book, the U.S. Center for Disease Control indicates that 1 in 44 individuals is diagnosed with an autism spectrum disorder (2022). The prevalence of autism continues to rise, as in the United States in the 1980s and 1990s approximately 1 in 2500 people was diagnosed with autism. Autism is the fastest-growing developmental disability. Autism is more prevalent in boys with approximately 5 to 1 boys to girls being diagnosed. Historically, girls exhibit variations on characteristic aspects of autism, and research supports the likelihood that they are under-represented in the current prevalence rates. There is also research that supports an under-representation of autism in the Hispanic and Black communities, with new efforts to connect and communicate more with those groups, to try and increase early and accurate diagnosis.

Characteristics

The three main aspects of autism are social, communication, and behavioral differences. Other parts of autism show up as challenges with executive functioning (which includes planning, organization, and time management), restricted or limited interests, and rigidity around schedules and structure. Autism presents as a spectrum, which means there is a range and varying scope of which areas of living may require more support or may be more prevalent in their presentation for the individual.

People with autism may communicate differently or exhibit behaviors that are distinctive, or unique, and unfamiliar to you. They may have varied language expression or be non-verbal. Some will utilize technology, cards, or signs to converse or indicate a need or want. A person with autism may display poor eye contact, unexpected actions that might include self-stimulating or soothing behaviors that could include spinning, flapping, or picking. They may have a low threshold for sensory stimulation and exhibit what is viewed as a 'meltdown' but it is caused by excess stimulation and can become sensory overload, which triggers a behavior.

It is possible a person with autism has a sensory hypersensitivity or hyposensitivity, which means an individual might be over- or understimulated by certain sensory experiences. Some sensory experiences that a person with autism might have difficulty with include crowds, loud sounds, small spaces, textures, lighting, certain smells, visual stimuli, touch and tactile sensation, taste, temperature, and unexpected sensory experiences, such as an alarm that may go off without warning. Some examples of how these sensory-related behaviors can manifest are that an individual may only wear clothes that are a specific material and that don't have tags, a haircut can be a challenging experience, they may wear headphones to block out noise, or they may have difficulty eating foods with certain textures.

Individualities that people with autism may experience include a challenge to make friends, a preference for routine and constancy, partiality for predictability, and social challenges that can include things like initiating or ending conversation. When understanding behavior, it is important to understand that behavior is communication, which means that when a person with autism is demonstrating a behavior, they are communicating a need, want, or response to stimuli. It could be that a person is needing something tangible or physical and they are reaching for it, or that they may try to escape or avoid a particular situation where they may be experiencing too much sensory stimulation, a person may display a certain type of self-soothing behavior that they utilize to help them feel more comfortable due to stimuli that is beyond their control, which is impacting them in a certain way. Those behaviors will manifest in ways unique to an individual and may or may not be particularly observable.

As a business, providing support for a behavioral situation is dependent on the situation. For example, if you are working one on one with an adult and you observe that they are displaying a behavior, or that the frequency of the behavior has increased (such as clicking the end of a pen cap), it is useful to stop the conversation and ask if there is anything that they are worried or anxious about, or if there is something in the environment that can be changed, like turning a fan off or reworking the lighting. If you are planning an event, have a low-sensory or quiet space available for people that may need a moment apart from a crowd or loud noises, and can be supportive. If you experience a parent who has a child exhibiting a behavior can help to ask the parent if there is anything you can do to support them, and to recognize that parents often endure a lot of judgment in public when this happens, and to support them in having the space and understanding they need in that moment.

Some autistic people experience repetitive and restrictive behaviors. Repetitive behaviors can manifest in many ways and may or may not be obvious to an observer. These could be physically repeating behaviors such as rubbing the edge of a shirt sleeve, or larger behaviors such as always brushing teeth the same way with a similar motion, angle, and process. Some repetitive behaviors can increase or decrease and may have meaning. There may be a time that an increase correlates with intensified anxiety or stress. It is important to not overly interpret these behaviors, but to be aware of them and to ask questions about their meaning when possible. Sometimes neurodiverse people are stigmatized for these behaviors, which is why it takes an increased personal awareness to not judge or determine the meaning of behaviors, but to understand that they may be a form of communication or coping mechanism.

There are some autistic individuals that demonstrate restricted interests. These interests may present as a preference, such as thoroughly enjoying a hobby or activity, or they can show up for a person as an area that can constrain their daily activities. For example, a restricted interest around food can impede balanced nutrition. A restricted interest about an activity might not allow a person to experience a wide array of options and be limiting. As a business, this may be an area to address. For example, an individual with a restricted interest that goes to a theme park may want to ride the same ride over and over again. The theme park will need to consider how that restriction can be accommodated for when they create neurodiverse accommodations and supports.

Difference versus Deficit

I'd like to take a moment in this section to discuss the concept of difference versus deficit. In medical terminology, and what has become a societal construct, people with disabilities are frequently seen as those that have a deficit in an area. As human rights have become more and more front and center, it is important to acknowledge that difference does not mean deficit. Over my many years of working with individuals with autism, especially the teen and young adult population, it is clear to me that we all have a unique way of being, and that does not necessarily prevent any person from reaching their goals and dreams. While I absolutely support evidence-based practices and therapies to help both children and adults with autism develop skills to navigate the world as it is, I also believe that as a society, we need to do better to shift the lens of 'difference' and to celebrate the unique parts of what each of us brings to our communities.

Neurodiversity

A term that has come into the conversation more recently, is neurodiversity. In the past, we've considered diversity from a cultural and ethnic perspective. Neurodiversity is looking at people with varying ways of processing the world around them. It may be that they are autistic, or have a different developmental disability. It could be that their intellectual abilities are different than the general population or that they're communication style is not typical. Is important as we look at neurodiversity to recognize that something that might be considered a deficit by some is only in fact a difference.

Hard Conversations

What happens when you feel uncomfortable, or you don't know what to say or do, when you encounter a person with autism? Please understand that it is okay to have those feelings. These are concerns that have been brought to me while facilitating a training about autism, or I've had someone pull me aside privately to ask this kind of question, because they might feel a little ashamed for their feelings. The most important thing to remember is not to rush to judgment, and to ask questions. For example, when you encounter somebody who communicates differently than you, it can be frustrating or challenging to understand their needs, especially when you are trying to help. What you can do is ask questions that aid your understanding. Ask the person about their communication methods, or try to communicate using a different method yourself. If you encounter a parent having a difficult moment with their child, ask them how you can help or if there is anything they need.

*Pre/Post-Test Quiz Answers

1. C
2. B
3. B
4. A
5. a. Social b. Communication c. Behavior
6. B
7. B

Chapter 5

Training

Information is power. The chapter on autism offered in this book may not provide enough detailed information for certain employees on your team, as it is designed to be a brief overview. More in-depth training is especially important for staff working in positions with responsibilities such as child care workers, law enforcement officers, security workers, and direct service personnel who will have unique and more individualized contact with people with autism. This chapter will guide the reader on what to look for in a training provider and how to match that to the needs of your business. This chapter will also discuss how to obtain training, how to deliver it to your staff, and what is needed after the initial training to continue the conversation.

What Training Does My Team Need?

When considering training needs, it helps to recognize that there are two main types of employees. There are those that are customer-facing, and those that are not. For employees that are not customer-facing, a basic overview of autism is useful to ensure they are included in the Autism Friendly initiative.

DOI: 10.4324/9781003185406-8

While they may not work directly with the public, they will certainly have neurodiverse colleagues that they interact and collaborate with. For employees that are working directly with customers, a more in-depth training is needed. How in-depth depends on the role of the staff member. A person who works the front desk at business, who is a greeter, or a customer service representative, is likely to encounter many people with autism and related disabilities, but is not likely to work very closely with them over a long period of time. An employee in the role of a coordinator, that is providing a direct program, such as an arts and craft specialist, a travel agent, an accountant, or a realtor, they are more likely to have sustained direct interactive contact with autistic customers. Those team members can begin with a simple training, but they will require a secondary or more expansive training that includes strategies on how to work closely with the customer they are serving in specific situations. When consulting with a trainer or agency, it will be important to share with them the role of the primary employees being trained, as well as any concerns or potential issues may have come up in the past that need to be addressed. This way the trainer can be prepared to provide the level of education that is needed and to address and individualize the training for the needs of your business.

If you are feeling apprehensive about engaging consultant for this initiative, or are not sure exactly what level of training your staff needs, I've provided guidance in the form of a questionnaire to help with the process. You want to begin with a foundational autism overview training, and from there, you can conduct a brief survey of your employees or a review of their roles, to determine their needs as well as their gaps of knowledge in this area. When initially seeking a trainer, look to your local community autism or disability-specific agencies and organizations, as they will not only have the content areas expertise needed, but also a strong grasp of the aspects of your region that are unique to the customers that you serve.

*Worksheet: Questions (and Answers) for Potential Training Providers and Consultants

- What is your expertise in the area of autism and neurodiversity?
  - You are seeking someone who has worked directly with autistic people for at least three years.
- What is your educational background?
  - Most professionals that work with neurodiverse populations have a college degree or professional certificate. Degrees may be in Autism, Special Education, Exceptional Student Education, Speech Language Pathology, Social Work, Psychology, Disability Studies, Vocational Rehabilitation, and Human Services.
- Are you certified or licensed in an area related to autism, neurodiversity, or disabilities?
  - There are two main certifications for professionals that provide behavior analytic services. The graduate-level certificate is a Board Certified Behavior Analyst (BCBA), and the paraprofessional certificate is a Registered Behavior Technician (RBT). There are also autism certificate programs offered by colleges and graduate schools. Other related professionals may have certifications or licenses in their area of practice such as a Licensed Clinical Social Worker, Licensed Psychologist, Speech Language Pathologist, and Certified Rehabilitation Counselor.
- Will a person with autism be present for the training or contribute to the development of the training?
  - It is important to have a person with autism contribute to either the development of the training or be present during the training.

- What are the topic and the title of the training?
  - Titles may range from autism overviews, to autism basics, to autism friendly strategies, and neurodiverse accommodations and supports.
- Is the training in-person or online?
  - Online trainings allow for many attendees who are in various geographical locations, but can have a less personal feel. In-person trainings offer more opportunities for questions and conversation.
- How long is the training?
  - An autism overview training should be around 90 minutes.
- How many people from your team will be providing the training?
  - Some organizations offer one trainer, which may be sufficient. Additional experts can be useful when taking a deeper dive into autism.
- Is there a limit on the size of the audience for a training?
  - Audience size can impact discussion opportunities and opportunities for informal exchanges. Some trainers prefer no more than 25 attendees while others are comfortable training hundreds of people at a time.
- Is there a cost for the training?
  - Cost will vary based on expertise, location, development time, and effort required to deliver the training.
- Can the training be offered in an online recorded format?
  - A solution for large businesses can be a recorded training that is added to the onboarding process or annual education series. For these types of trainings, it is recommended to embed a post-test to help ensure that information was received and understood.
- If my team requires a more in-depth training or a follow-up training, is that something you or your agency is able to provide?
  - Some organizations have a plethora of experts and are able to provide education in areas such as autism and

safety, employment, mentoring, assistive technology, and other topics that may be useful to your business.

- Is there any technology, or materials that we would need to provide as part of the training?
  - Typically, trainers require options for visual presentations which may include videos and slides.
- What is your availability? Are you able to train our team before or after typical work hours? Or possibly during their lunch break as a lunch and learn?
  - For some businesses, adding extra time to an employee's day is outside of the reach of the company's budget. Offering training as a lunch and learn is an option to incorporate education into the work day.
- Is there anything we need to do to prepare our team for your training?
  - It is important to share your vision with your employees before bringing them into a training room. Part of enrolment is your own team. A strong trainer can help you with this process as preparation for your team.

How to Continue the Conversation

A 90-minute autism overview training is not the end-all and be-all of the education portion of becoming an Autism Friendly Business. It is necessary to continue the process of educating staff, and supporting their needs as your business adds divisions, programs, expands services, and implements change. It is also important to be sure that new staff receive Autism Friendly training during the onboarding process. Many businesses will simply incorporate an inclusion and/or disability aspect to the new hire orientation trainings and either bring in professionals to offer that training, or be trained themselves in a train-the-trainer type model.

As staff interact more with people with autism, you may consider encouraging them to make note of things that they need further training on. For example, if a sports coach is working with a child who exhibits some behaviors during a practice, that coach may want to indicate to the inclusion lead that they need a training on behavioral strategies. Other times to provide continuous training is as programs cycle through staff. For example, summer camp staff tend to be hired on an annual basis and the new team will need training each year. You may also have program specialists that come in to provide just one service. An example may be an artist that comes into your organization to teach one particular painting class. This person would not be someone who is full-time staff, and likely would not have received this training. It is helpful to incorporate at least the basics of these trainings into the hiring process even for very part-time workers. A training calendar and strategy can be developed by either the inclusion lead, the ADA specialist, or your human resources department.

Chapter 6

Create Your Inclusion Team: Community Partners Are Key

To be an effective Autism Friendly Business, you will benefit from establishing an Inclusion Team. One of the most valuable ways to increase neurodiversity is to bring together your local experts. Not only will this group become an advisory support for your company, but this is an excellent way to develop relationships with community agencies that work with people with autism. This chapter will discuss who to include in your team, what types of organizations will be helpful to consult with you, how to bring those people together, a sample meeting agenda, and the best way to utilize their expertise to inform your business.

Develop Your Inclusion Team

The process of assembling an Inclusion Team typically begins with your ADA coordinator, outreach professional, or human resource specialist taking the lead and reaching out to local autism and disability service agencies and non-profit organizations. Directly contacting each individual entity can be

DOI: 10.4324/9781003185406-9

time-consuming, and a more efficient way to begin exploring the potential support that is available in your region is to seek out an established work group, task force, or committee, where these organizations meet together. The group may give you the opportunity to present your ideas and describe your needs to a larger number of agencies that can then determine if they have the time and resources to participate. If you'd like to have a more in-depth conversation before reaching out to a work group, or you don't know how to locate such a group, then I suggest you reach out directly to a single organization and start the process one-to-one, meeting together to brainstorm a partnership and learn about what already exists in your community. It is important to start creating a list of people, agencies, organizations, and non-profits that may be a good fit for the goals of your business, for which you can develop community partnerships. From there, hopefully, that individual will have connections across the disability community and can help you broaden the scope. Many of these agencies and non-profits are already part of community task forces and committees, and it is likely, that if you can locate one of these groups, you may have access to multiple professionals that can assist you. For example, in my region there is a Special Needs Advisory Council, an Interagency Coalition, and a Transition Task Force, amongst others.

When gathering a group together to be your Inclusion Team, you will need a variety of perspectives that support and represent the different aspects of what you are trying to accomplish. People will have different reasons for signing onto your project, but all of those reasons are important. The teams I have participated on included representatives from organizations that work with people with autism, those who have an intellectual disability, people who are blind, deaf, dyslexic, or have cerebral palsy, amongst others. You want to be sure to include people with autism, at least one parent, an educator, and organizational representatives that have professional expertise which includes direct service. Be sure to also consider gender, culture, age, and socioeconomic diversity in your group.

Depending on your company's structure, it may be prudent to work with your community leaders, who may be politicians. You may discover that you need to present your initiative and be prepared to explain the return on investment and potential use of resources. Education will also be needed to help your leaders with understanding and learning more about people with autism. There are many leaders that support these types of inclusionary initiatives, but may not be willing to spend the time or money needed to do it well. If you find that is the case, it may be useful to connect with a business who has become Autism Friendly and share their model and success with your leadership team.

Titles for this type of group can be straightforward or creative, and may change as the purpose of the group becomes clear or the goals develop. For example, I've known groups to be called the following: The Inclusion Support Team, the ADA Consulting Group, the Inclusion Consulting Team, Special Needs Work Group, or the Special Needs Advisory Team. A good approach is to invite the potential participants to the table and let them work together during the first meeting to name the team.

Brainstorm and utilize the following worksheet: Create Your Inclusion Team. Consider who might be useful from both inside and outside of your region. It is important to include at least one individual with autism on your inclusion team. This person would ideally be representative of the people that participate in your programs and can provide insight and feedback. Ideally, you will have representation from more than one age group. Some teams include an adult with autism, and a parent of a child with autism, to ensure that at least two different perspectives on needs and engagement are considered.

*Worksheet: Create Your Inclusion Team

Potential Employees as Team Members

Name	*Department*	*Job Title*	*Reason to Include*	*Contact Information*
C. Bell	Communications	Brand Developer	Writes content for social media	c.bell@haworth.org

Potential Community-Based Team Members

Name	*Organization*	*Job Title*	*Reason to Include*	*Contact Information*
J. Specialist	Autism Center	Clinician	Provide autism expertise	j.specialist@asdctr.org

Location and Timeframe for Meetings

Most Inclusion Teams will either meet quarterly, or once a month. Monthly is typically best, depending on your business and frequency of projects or programs occurring throughout the year. It is important to recognize that the people on your teams may have certain seasons that are busier for them, or times they are not in the office, such as teachers or educators that go on summer break. When designing your meeting structure and calendar, be sure to consider location. Some Inclusion Teams will always meet in the same conference room at the business that they are supporting. Other teams will rotate the specific location so they have the opportunity to tour and view various buildings and different presentations of the same business. For example, a retailer may have a store in a mall and stores in a retail center. The mall is less likely to have immediate safety issues with cars passing by the front door, where a shopping center location may have a front door that leads directly out to a street or parking lot. Some teams will meet at their members' locations so that they have the opportunity to learn about the local services being offered and to deepen those relationships. Throughout COVID, many groups met online, which allowed for more attendees, but not for the knowledge and community building that comes with in-person experiences.

A good timeframe for meetings is approximately an hour and a half. One hour tends to be a bit short, and two hours can be a long time for people to hold focus. There may be opportunities to extend a meeting if the group is also touring a facility or program as part of the gathering. Other reasons to lengthen a meeting is if your larger team has smaller work groups that require time to convene and discuss a specialty area, project, or event they are planning. Options for additional meetings can include inviting your members to train the group about their organization or population, or to include the team in planning meetings where they can provide expertise and contribute.

Inclusion Team Meeting Agenda

During your first inclusion team meeting it is essential to orient the team to your business and each other. When designing a meeting agenda for your Inclusion Team, it is important to include time at the beginning for introductions and what we call agency updates. This is an opportunity for each agency or program or education representative to be able to share the things going on in their universe. Perhaps it is autism awareness month, or possibly there is a fundraiser or conference be offered by one of the representatives. It can be a forum for these groups to share information with each other and see where there may be some over labs or opportunities for partnerships. This is also an important time for the business to hear what's going on in their local community, and to see where they might be able to plug-in or support. Perhaps an agency is having an outreach event where they are providing tables for different groups to share their materials. The business, for example, might want to bring some of their staff to an event as a sponsor or to provide information, and begin to really connect with and invite members of the disability community to engage with you.

One of the most important aspects of the inclusion consultative group is to provide input and feedback on upcoming events, projects, and strategies. This is going to be the core of the meeting where the expertise of this group will be guiding the content areas of your initiative. After agency and program updates it is useful for your business to provide a brief of where they are on their autism-specific or disability-specific projects. For example, you may be creating a social story for children whose first time it is to travel on an airplane, or visit your amusement park. It is useful to share these things with the group as they can add evidence-based aspects that are geared to develop a quality and successful product. Your business may also be hosting an event during which the inclusion representatives can participate where they can share their disability service-related information with customers. For example, a ballpark might have a baseball game that benefits local autism organizations and they could offer agencies a chance to conduct outreach efforts to the attendees that day.

Outside of regular inclusion consultative meetings, it may be prudent to invite members of that committee to provide expertise on specific projects. For example, when the city of Boynton Beach was designing their downtown remodeling project, I was invited several times to speak with the architects and city planners about building design and safety concerns. These discussions included ways to corporate visuals and physical barriers to help create safe spaces for children with autism. Things like wayfinding and low sensory spaces were also discussed. When remodeling existing environments, it is also important to bring in the inclusion team to offer input and suggestions.

Sample Inclusion Team Meeting Agenda

INCLUSION TEAM MEETING AGENDA
DATE
LOCATION

I. INTRODUCTIONS
II. AGENCY/ORGANIZATION UPDATES AND UPCOMING EVENTS
III. BUSINESS GENERAL UPDATE
IV. EVENTS OR CELEBRATIONS
V. PROJECTS THAT COULD BENEFIT FROM TEAM INPUT
VI. REVIEW OF PREVIOUS PROJECTS/FOLLOW-UP
VII. FUNDRAISING/SPONSORSHIP/ANY FINANCIAL REVIEWS
VIII. TEAM OR PROJECT GOALS
IX. COLLABORATIVE EVENT PLANNING
X. PROPOSED NEXT MEETING DATE, TIME, AND LOCATION

*CAPTURE ATTENDANCE AND UPDATE CONTACT INFORMATION ON SIGN-IN

Team Collaborations and Projects

Collaborations are win-win opportunities for your business and Inclusion Team members. It is likely that members of your team bring ideas that are founded in their knowledge of autism or their disability area, and may be instrumental in expanding your reach. This can occur as participants will have different specialty areas. For example, someone who works with the Deaf population may be interested in training your employees in basic sign language and communication techniques, while a blind specialist may have contributions about wayfinding.

Other collaborations might include options for having your business participate in or sponsor an organization's work, event, or program. For example, an agency that provides behavioral therapy for children might be hosting a day where the police department brings out their canine unit and the therapist works on animal familiarity with a particular child or group of children. If you own a pet shop, are a Veterinarian, or manage a pet supply store, sponsoring and participating in this type of event would be a natural fit.

One example of a partnership was at the City of Boynton Beach, where they utilized their connections on their inclusion team to collaborate with a middle school and create a program for Arbor Day. The city park manager and a teacher created project for students and city staff to plant trees. This was very successful in bringing together students with disabilities and the parks department at the city. A multi-year collaboration I was directly involved with, with the city of Boynton Beach, because of my participation on their inclusion team, was several iterations of a developing Art and Autism Expo. These expos brought together artists that were autistic with the team at the city's cultural center. For one event, the artists were able to display their art throughout the center's public spaces wing, and were the official 'Art in Public Places' group for the summer. That event culminated in a Breaking Barriers reception and opportunity for the artists to sell their work and connect with the community. Another art expo was held during COVID times and was a hybrid event where artists were able to display their art at the city as well as in an online museum. These events provided incredible opportunities for the artists as they were acknowledged and given the chance to share their talents. It was also extremely beneficial for the city, as their art in public places program and cultural center was able to expand their level of inclusion by connecting with people with autism and bringing them into city spaces. A poster from one of the art events is shared below.

Regional or Statewide Inclusion Teams

Once your Inclusion Team is established, there might be an opportunity to expand the team outside from a central location to multiple regions. Some teams might grow into large work groups that include several cities, states, or even multiple countries. This can be beneficial for a global view because there is an opportunity to share ideas and to collaborate on larger projects. It is also recommended, however, to continue to have the initial central Inclusion Team that provides perspective on the growth and development of your company's inclusive strategies.

3 THE WHAT: CREATING CHANGE

You have the power to create change. As with so many initiatives and grassroots movements, they begin with the passion and motivation of one person. It is up to you to be that one person. As a society, we tend to be very outcome-driven, but to truly make a systemic difference, we must look at the reasons, the 'why' while we explore 'what' needs to be done. When Simon Sinek talks about his golden circle, he reminds us that organizations need to start with their 'why' and to couch their process in their purpose. This section will help your business explore the types of strategies you can implement to become Autism Friendly and inclusive, while developing those ideas through a lens that holds true to your core mission and what is at the heart of your 'why.'

DOI: 10.4324/9781003185406-10

Chapter 7

Getting Started

What is inclusion? Inclusion is not simply acceptance, it is a framework that goes well beyond just knowing about or acknowledging differences. Inclusion is a welcoming celebration and recognition of what makes each one of us unique, and valuing that as a contribution to society. It is honoring how our distinctiveness brings depth and layers to our communities and develops societal diversity. Chicken soup would be plain and flavorless if it contained only chicken. We need celery, carrots, onions, rosemary, and thyme, a diversification of ingredients and spices to create something delicious. As we focus on developing neurodiversity in your business, this chapter will help you identify the ingredients you have and the ones you need to get started.

DOI: 10.4324/9781003185406-11

*Worksheet: What Does Inclusion Look Like for Your Business?

Let's begin by asking questions and creating a dialogue. The following prompts will help your organization define inclusion and improve the lens of accessibility as it specifically relates to your business practices.

- Share five words that describe inclusion:
  ____
  ____
  ____
  ____
  ____

- What programs and initiatives have you already created that support inclusion?

- What is unique about your product or service that you will need to be aware of when creating an inclusive experience for neurodiverse customers?

- What barriers to inclusion do you currently experience, and what do you anticipate will be challenging areas for your customers?

- Does your business communicate with your customers in a variety of ways, and is that communication utilizing universal design practices?

- Does your business require that customers know or understand language that differs from standard definitions, or that is specifically unique to your product or service?

- In a perfect world, with no financial barriers or resource restrictions, what would your ideal inclusive initiative look like for your company?

How did it feel to answer these questions? Were there any that made you uncomfortable or worried about what your company will need to learn in order to be successful in this arena? A strong foundation starts with the commitment to change, which is why you are here. Commend yourself for that and after you are a few months into the process return to these questions and see where you have grown.

Chapter 8

Environmental Modifications

This section will explore the environment in your business and how small modifications can be made to remove barriers and enhance safety. A deep dive into this area includes examining physical space in buildings, outdoor areas, safety practices, creating sensory spaces, and developing accessibility strategies. Guidance will also be offered on topics like designing a large public event, and strategies will be discussed on topics that include how to make small changes that are inexpensive without sacrificing space.

When we design sensory friendly spaces, we begin by considering all of our senses and how autistic individuals may be impacted. Traditionally there are five senses, but we are going to expand that a bit and discuss those that might be an issue for a person with autism, and adding methods for a business to make small modifications that help neutralize the issues.

DOI: 10.4324/9781003185406-12

Sight

Our eyes tell us information about what we see in our surroundings. They are also are receptors of light. For people with autism, they may experience sensitivity to certain types of lighting, or levels of lighting. Specifically, high-intensity UV lights from above can be overwhelming. Lamps and dimmable lights can be a much better option. They can cause headaches and emit sounds that are distracting. Autistic individuals may have difficulty holding eye contact, maintaining a directional gaze, and visual input. When creating spaces, including picking lighting, colors, and design, consider possible visual hypersensitivity and the impact your choices may have on a person in your environment.

Hearing

The ears bring in and interpret sound waves. An autistic person may be sensitive to sound and may experience anxiety or a behavior change if sound is interpreted as loud for them. A sound that is loud for someone who is not autistic may be heard as loud for someone who is. You may see people with autism wearing headphones or earplugs in order to help them modify the sounds so that it works best for their own quality of life. When creating environments, it is important to consider how much sound we are pumping into the background. I encourage you to sit quietly in your space for five minutes and listen for ambient noise that you may typically not focus on because you filter it out. These types of noises can include sounds from air conditioners or lights, possibly background music, or other machinery like an elevator motor or ambient noise from people walking by. When picking a location for a quiet area or low sensory space, it is important to consider what sounds will be occurring in the room.

Touch

Individuals with autism may have some sensitivity to touch. This may be increased when the person they are experiencing touch with someone unfamiliar to them. For example, it may mean that it is difficult to shake a hand for a job interview, or to receive a hug at the end of camp from a counselor. It is important that this is not seen as a sign of rejection and not taken personally. Some individuals will have touch sensitivity around clothing. They may be most comfortable in certain types of clothes, and require that there are no tags in an item. This can be challenging if the business is offering a sports program where a particular uniform is required. These are places where modifications can be made by working with types of fabrics and styles.

Smell

Smells can permeate a space very easily. From cleaning supplies to odors that come from kitchens, smells are everywhere. When ordering supplies to clean with, it can be helpful to go with neutral scents or fragrance-free. This is especially important if the business is offering things like towels at a pool program that get washed in industrial laundry detergent. For programs where people come together in small spaces, it can be useful to request that people do not wear strong perfumes or body sprays in order to help keep the environment as fragrance-free as possible.

Taste

It is likely that a business that is not a restaurant may offer food items during a celebration, an event, or a children's program.

For these situations, it is useful to provide foods that are varying and type and texture. For example, if tacos are being offered having an option for a soft taco and a hard taco would give an individual a choice for texture. When working one on one in a food-related setting, it is best to consult with individual or family beforehand to determine likes and dislikes, and, when possible, offer choices.

Temperature

Temperature can be a big concern for a person with autism. For some, they may have problems experiencing sweat on their bodies. Others may have physical response to extreme cold. When having outdoor events, it is important to incorporate temperature regulation into the quiet space or sensory friendly space. In hot climates, it can be useful to have these spaces in air-conditioning, or at least in a shaded area. In colder climates, having warming options such as heaters or blankets can be useful.

Spatial

There are some individuals with autism that have challenges with spatial relations. This may mean that they have a difficult time judging distance and direction. Utilizing visual supports that mark sides of path, or creating maps and wayfinding strategies can help support these individuals. For safety, add barriers to prevent individuals from accessing water or busy streets.

Sensory Spaces

When designing spaces, you may hear about a 'sensory friendly' space. That phrase is very general as environments can be low sensory space or high sensory space. It is important to differentiate between the two both for yourself as you engage with the design process, and for your families. If an autistic person is overstimulated in a program or event, and they enter a sensory friendly zone in order to have a break from the stimulation, they may have further difficulty if they walk into a room full of high sensory objects. A low sensory space will be the opposite and is typically a quiet room with low lighting, minimal to no sound, soft furniture, and walls with little to no distraction. While creating a high sensory space, you may incorporate bolder colors, tactile experiences, ways to move the body or engage with other types of motor skill experiences, and the physical space may be more out in the open. There is an example of both types of space in Chapter 7, which is about developing these types of environments in a library setting.

*Worksheet: Design Your Space

This worksheet will guide you through modifying a physical space or environment at your business:

- Is the objective of the environment to be high sensory or low sensory?
- How will the environment serve customers? What is the purpose?
- What items do you currently have that can be used or modified to create a new space with this function? (i.e. furniture)
- What items do you need? (Fidgets, high tactile boards, blankets, mats)
- Does the space require painting?
- Is the area easily accessible to those it is intended for?
- Where is the door to the room? Does it lead to a parking lot or is the exit near water? Can an alarm and visual support be installed on the door?
- What other safety modifications are needed? (i.e. Does the space have outlet covers? Are items, like a tv, mounted to the wall?)
- What types of visual supports and wayfinding can you create to help people locate this space?

Chapter 9

Communication Techniques

This chapter will offer you strategies and suggestions on how to improve your communication approach when working with individuals with autism. These supports will include verbal communication, email, print materials, and marketing. Communication is how we share ideas. When we provide information to the customers in our communities, they are able to make informed decisions related to how they live, how they engage, and the way in which they're able to access their needs. Communication occurs in many forums, including: through the internet, social media, newsletters, text alerts, flyers and visuals, through voice, and in the environment. When we design these methods of communication, and spaces, it is essential that we consider the needs of people with autism and other disabilities. In this chapter, we will review effective ways to communicate with this population.

It is important to start off by helping making people feel included. Everyone wants to feel welcome. Offering an open-door policy for communication, as much as possible, is helpful. When that isn't an option, providing the public with

DOI: 10.4324/9781003185406-13

engagement surveys, contact points, or public meetings can help curb the frustration of not feeling heard. We want to have strong communication skills, which start with asking questions. Ask questions to customers with autism, and speak with families. Find out how you can be helpful and interact.

First Point of Contact

The first point of contact a customer has with your business is a decision point. This is when they will decide if they will engage or participate, or if it is too complicated, cumbersome, and that to have their need is inaccessible. That first point of contact can be many things. It might be your website, or your phone message, or a sign on your door or bulletin board, or the person sitting behind the desk. Each one of these methods of communication communicating information can be frequently overlooked as a place to include strategies and accommodations. We are going to review ways to improve and enhance your methods of communication in order to make them more inclusive.

Universal Design

In all forms of communication, we want to incorporate aspects of universal design. This type of approach will allow individuals that use a variety of methods of communication, to access and understand what is being shared. Many websites include some aspects of universal design, but if it is possible to bring this lens to all that you do, your customer will be more likely to engage. Accessibility starts with safety which can be eliminating flashes, removing blinking signage or icons, reducing the amount of colors, and adjusting colors to be easier for the eyes to see. To enhance the view, your user should be able to adjust the text size, graphic dimension or directions, adjust for a light or dark background, and possibly even change the font. Fonts are recommended to be simple and without excess ornamentations. Consider spacing, alignment, contract, hyperlink indicators, and animation. Design also includes considering reading level and language accessibility. Some people with disabilities will utilize technology in the form of a screen reader that reads the text on the screen and speaks it aloud to them. Making sure that your site is able to be used by these readers is essential.

Website

I suggest you take a look at the homepage of your website. Does it meet standards for universal design? By this, I mean does it begin to have accessibility options for people with disabilities? For example, is there a text-to-speech option included? Is any document posted to the site such as a newsletter or flyer, in a format that is usable by a screen reader? Are there options to modify font size? These are very basic technological changes that can be made within the coding and posting on your website to ensure at least very basic accessibility.

The next step is to evaluate how the environment of your website looks. Is it flashy, or busy? Is your website full of distractions? Is it difficult to locate items or areas that people might need information about? Are things hidden in drop-down boxes or under a series of lines or characters in corners they require multiple clicks to navigate? Is your website all text? Does it require a high level of literacy? Have you included any visual supports on your website? Do you have icons or graphics that represent areas to visit on your site? For example, if you have regular meetings that are open to the public when you share the calendar of meetings is there an icon that represents a calendar? If you have shipping information on your site, is there an icon of a box, which can help make it easily identifiable? Are your icons able to be clicked on for ease of navigation? Here are some examples of poor website designs that are complex, very difficult to understand, and have information hidden. These sites are busy and full of distractions that are likely to give anybody a challenge, but for a person with autism, who may have difficulty with the overstimulation through their senses, they are likely to have a poor experience with this type of website, and will move on without obtaining the information they were seeking.

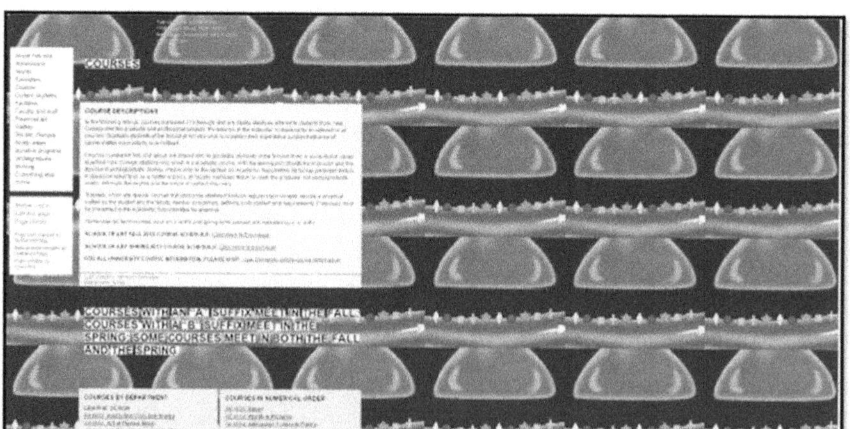

Here are some examples of websites that are designed for easy use, with accessible options, and with a high-level visual support, for those that are interacting with the site.

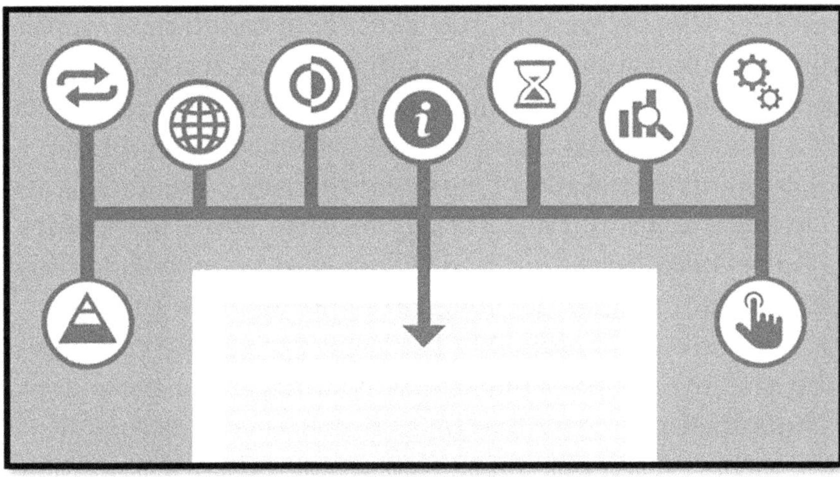

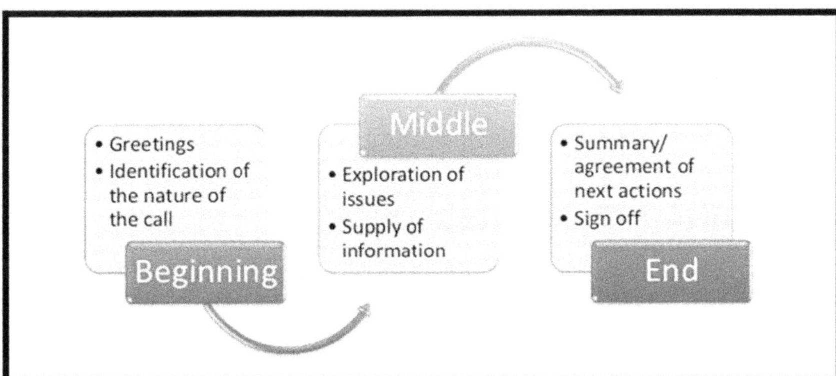

Most organizations and businesses frequently update and change sites as offers and programs come and go; however, it is important that we provide some consistency to users and it can be beneficial to update information in the same location on the page. For example, having one section that is a newsletter or flyer that frequently gets updated in the same space and with the same icon, as opposed to reworking all of the content on the page and moving things around. Our home pages may be in good shape, but after looking through those, consider your secondary pages, and what happens when someone clicks on an icon that takes them to the next website within your business' page. Does that remain accessible? For example, the person who clicks on the swimming icon, when they arrive at the page that provides information about the pool opening and closing times, and the rules, are all of a sudden, the icons and accessibility features gone? Sometimes this occurs because we have an IT specialist know a technology specialist put together our templates and we have program managers or coordinators that then will update the contents of the sites over time. While we look at design for these areas, it is important to ensure that there is consistency across all of the pages within our entire website, and then when information is uploaded by people that are not the technology specialist, they are provided with directions on format including what is usable by screen readers, and meets accessibility guidelines.

Using Direct Communication

When I teach organizations about being Autism Friendly, I challenge them to go an entire day without using a simile, metaphor, or whatever the heck those things are called, and to only use direct communication. This is more difficult than most people believe. We frequently use words that have multiple meetings. For someone who is on the autism spectrum, it may be difficult for them to understand which meaning of a word you are intending them to determine the meaning. The example of it's raining cats and dogs is too vague and confusing. A better statement is to say that there is a lot of rain coming down from the clouds.

One example I like to share is a word that can get us in very deep trouble if we are not specific about it. I had a wonderful meeting with a college-age student to discuss his career goals. After the meeting, I walked him down the hall towards the elevator. I said please keep in touch. He looked at me like I had three heads. I quickly recognized that he had misinterpreted the way I was intending to use the word touch. I quickly said to him, no no, I mean please email me your updates as you make progress towards your career goals, and I will email you back. We use these phrases that can be very easily misunderstood.

Unfortunately, this can lead to a person not understanding what you were trying to tell them, and having a lack of follow-through, or follow-up. There are times that this can be particularly detrimental to a person's life, especially if you were working with them on something important. I have seen professionals see the confusion in the eyes of an individual they are working with, and instead of rewording what they are trying to explain, they repeat the same phrase multiple times. Sometimes they even begin to raise their voice. This is not helpful. This is a situation where a statement must be rephrased with different language, and different word choice, using words that are exceptionally clear, and direct. Words that have singular meeting, or where the meeting can be explicitly explained.

These methods can help support clear communication:

- Do you allow walk-in appointments?
- What is your typical time to return voice messages? (24 hrs.?)
- What is your preferred method of communication?
- Explain what you consider urgent and what is not urgent.

Recognize that it will take time to build trust with the individuals you are working with. Some areas to be aware of are:

- Mutual respect
- Reliability

- Customer-centered
- Strong listening skills
- Validate experiences
- Be honest and genuine
- Being a safe space where a person feels free to share without judgment

One-on-One Communication

When we were examining first points of contact, many of those are considered indirect from your business, but are important moments to your customer. Indirect ways of connecting with customers can include things such as websites, phone messages, flyers, and newsletters. Direct contact is considered communication that is an exchange by more than one person and is a give-and-take of information. With direct contact, these are the times we are you are communicating specifically with the individual and not with the community at large. The most frequent times direct contact begins is when you receive either an email, a letter, a phone message, or an in person walk in. These will be your first likely points of direct contact. Frequently this direct contact is in the form of a person trying to obtain clarification of the sources of the indirect contact. It is important to note if you have any themes of your direct contact and see if you can resolve them with changing up either the continent or the method in which you deliver the general information. For example, if you receive ten phone calls a day asking when the swimming pool will be open, it is advisable to post those hours on your website or in the business spaces.

Let's examine a direct contact of an email. The email is likely to be sent to your business's general email box. In the United States, this might be an 'information at yourbusiness.com' type of email. It is important to note who is checking your general information email. For some businesses this may be done by volunteers, part-time employees, or people that are not in administrative or leadership roles. When you are training your staff, it is important to make sure that these people are included, because they are often the first point of direct contact.

When you receive an email from a customer that asks maybe one or two questions. The email may ask for directions to an event, or for more information on how to register for a program, or how to participate in a meeting. Let's use the example of the person who was asking for directions to an event. When responding to an email like this, it may be a copy-and-paste pre-written paragraph of text with directions. If this is a large paragraph of directions, it can be overwhelming to a person with autism. There are some simple steps to make this email more understandable, accessible, and less consuming. One of the things I suggest is that you start by breaking the directions down into a bulleted list. After separating out the directions into these smaller statements, and visually breaking them up to have them each on their own line in the email, this will help with understanding. If you are using a word processor, or even most smartphones, this is an easy use of the technology that we already have at hand that because of straightforward visual support. To go one step further, utilize the bullets that are open circles. The open circles transform the list to a checklist, which allows somebody to either print the list and then check off each bullet or direction as they complete them, or to check the boxes off on their phone or computer. Another option is to change your bulletin list into a numbered list. A number list can be useful because it provides some chronology. For example, it indicates what comes first and what comes next simply by using a set of numbers. While these ideas may appear to be quite simple, the difference for the user, the person who is going to be following these directions, can be significant. When items or tasks are broken down into small steps, and they can be marked as complete as the person is accomplishing them, it can make an experience that might be overwhelming or appear to be not accomplishable, a manageable task.

Another way to support one-on-one communication is to create an agenda template that can help an individual and yourself have a successful meeting. This is a document you would provide your customer and they can fill out before a meeting or during a meeting to help them prepare for a subsequent meeting. Types of businesses that may benefit from agendas or forms similar to this are realtors, hair salons, legal services, banks, doctor's offices, and interior designers, amongst others. This is useful because it shows that you support taking notes during a discussion, that there are prompts that include reminders to help the person be successful in bringing what is needed to the meeting, and that the goal of the meeting is more likely to be accomplished with preparation and planning.

*Meeting Agenda Template

This agenda template can be utilized by autistic individuals to organize and plan a meeting.

MY MEETING AGENDA

Meeting Date: ____________ Meeting Time: ____________

Meeting Location: ______________________________________

Who I am meeting with: _________________________________

Purpose of the meeting: ________________________________

What I need to bring with me to the meeting:

____

Questions I have for the meeting.

____

Items or information I need to obtain.

Date, Time, and Location of the next meeting:

____

Meeting Notes:

____

____

____

____

____

Chapter 10

Visual Supports

This chapter will explain and describe visual supports, and will discuss their origin and effectiveness. Samples will be included for multiple types of situations and settings.

When I conduct Autism Friendly trainings in person, I am usually accompanied by a cup of caffeine-infused deliciousness from one of the world's famous coffee houses. I am a huge supporter of their company, but I do tend to use them as an example of how important visual supports can be: Take a moment and imagine that you are a young adult with autism going into one of these coffee stores for the first time after receiving your first pay check. You are excited to be able to order a blended beverage with whip cream and chocolate drizzle on top. You would like to walk away with the largest one of these beverages that is available. You walk into the store and see stacks and stacks of cups. The person behind the register asks what size you'd like, and when you hesitate for a moment tells you the name of the three sizes. She says, tall, Grande, and Venti. Logically, you decide to get a tall. There is no picture of the cups and no indication of which one is which size. After lots of training sounds and excitement waiting for what you are proud to spend your hard-earned money on. When you are handed your beverage, the cup is tiny. You see

DOI: 10.4324/9781003185406-14

other people with larger cups. You ordered a tall. You thought that was the tallest cup they offered, because of course it would make sense that a tall is tall. However, this is not the case. Why would the tall cup be tiny? You are confused and frustrated. Your excitement is gone. You walk out of the store with your small blended beverage, disheartened, and probably will not go back into that location again.

What would help our coffee buyer? A visual support. A visual support is a very simple way of providing information beyond the use of words. All this coffee company would need to do would be to have pictures of the different cups on a board and have the names of the sizes underneath those cups. A simpler and cost-free way to do this would be to write the name of the size on the actual cup and put each one on the counter as a visual display. Sometimes such a simple act, such an easy modification provides an accommodation and improves the experience and access for a person with autism.

There are a few different types of visual supports. Some visual supports are simply one shape like a stop sign. This indicates to the person to not go further, not to open a door, or to stop a particular behavior. Other visual supports might be a little more complex. They might have two or three different items on them indicating what something is. While other visual supports are incorporated into a larger word-based item such as a restaurant menu, where the visuals are on each side of the food description. Some visual supports provide simple directions, such as three steps to withdrawing money at the bank, or three steps to set up for an art project. Visual supports can also come in the form of a task list. This can be done by creating simple directions and having the design so that the person can check off or ask out the direction once it's completed. Visual supports can be easily developed on a computer and printed out, posted, and updated as needed. The following samples show different types of visual supports that were created for use in real-life situations.

*Sample Visual Support Template

How to:

3
EASY
STEPS

Step 1:

Step 2:

Step 3:

You are finished!

CITY OF BOYNTON BEACH

www.Boynton-Beach.org

FAU
CENTER FOR AUTISM AND
RELATED DISABILITIES

*Sample Visual Support Incorporating Directions

I Get Ready to Paint

Step 1: I put on an apron to keep my clothes clean.

Step 2: I sit down on a chair at the painting table.

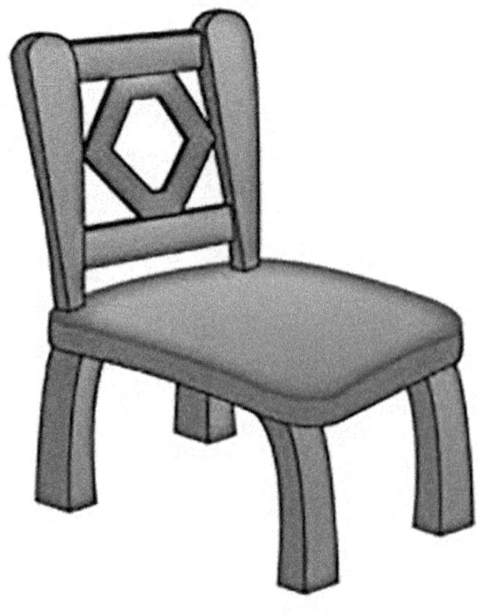

Step 3: I wait for instructions before I start to paint.

I Clean Up After Painting

Step 1: I put my used paint brush into my cup of water.

Step 2: I take my apron off and put it away.

Step 3: I go to the sink and wash my hands.

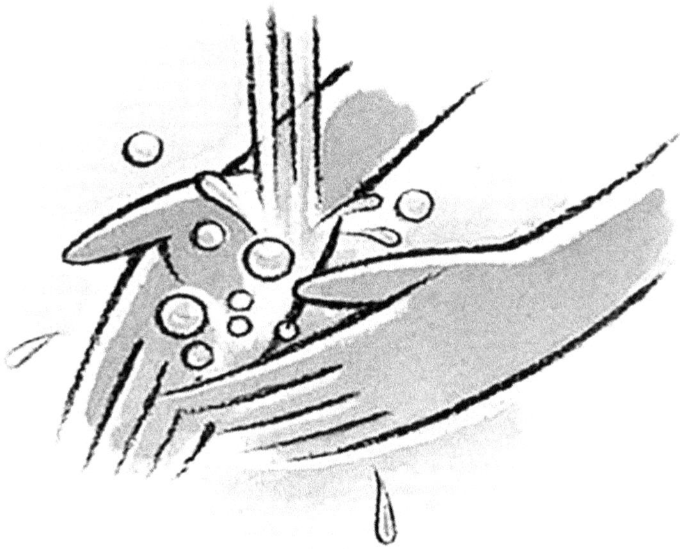

*Sample Visual Support Where Graphics Are Incorporated into an Existing Document

Restaurant Menu

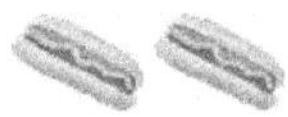

2 Hot Dogs **$2.99**

Toppings Include: Ketchup, Mayo, Relish, Chili or Sauerkraut, Onion, Cheese

2 Tacos (choice of beef, chicken or pork) **$3.49**

Toppings Include: Lettuce, Cheese, Sour Cream and Salsa, Tomatoes

2 Veggie Tacos **$2.49**

Choose from: Lettuce, Onion, Peppers, Olives, Sour Cream or Cheese, Tomatoes

Burrito(choice of beef, chicken or pork) **$4.99**

Toppings Include: Lettuce, Tomato, Black beans, Sour Cream and Salsa, Rice

Quesadilla with Cheese (choice of beef, chicken or pork) **$4.99**

Sour Cream and Salsa served on the side

Grilled Cheese w/Bacon Combo **$5.99**

(Choice of American, Swiss or Provolone) Served with chips and a drink.

Wraps (Choice of any Gourmet Hero) **$5.99**

Italian, Chicken Salad, Tuna Salad, Turkey and Bacon, Veggie, Corned Beef

Spaghetti and Meatballs **$7.99**

Served with a small house salad (lettuce, tomato, cucumber, choice of dressing)

Additional Toppings

 Lettuce

 Black Olives

 Tomatoes

 Cucumbers

 Onion

 Carrots

 Cherry Pepper

 Alfalfa Sprouts

 Jalapenos

 Avocado Spread

Chapter 11

Social Stories

Why is it important for a business to create social stories? There are many opportunities to use this communication aid to help improve accessibility for people with autism. Businesses can use social stories to help customers navigate new experiences, different environments, learning directions, or becoming acclimated to a potentially uncomfortable situation. In this section we will review what a social story is and provide directions on how to create one, offering samples to use as a template.

What Is a Social Story?

Social stories were designed by a teacher named Carrol Gray in the early 1990s. They are used to communicate information and reduce anxiety about situations that an autistic person may experience. Social stories are highly visual, and utilize first-person language, from the reader's viewpoint. For example, one may start by saying 'I am going to the park today.' The tense should be present or future, depending on when the activity or experience is taking place, and where the activity is located. It is best practice for a social story to be made available before the activity. The family will read the story together and

DOI: 10.4324/9781003185406-15

the child will have a good idea of what the activity is going to entail, including places where there may be rules, or potential challenges. If the social story is only being made available in the physical location where the event is occurring, for example, 'today we are learning how to float in the pool,' and the social story is being made available at the pool location, and being read by the swim coach, then it is likely to be very much in present tense.

What Must a Social Story Contain?

Social stories need to contain certain components. They include descriptive, directive, perspective, affirmative, cooperative, control, and partial sentences. It is important to Emphasize the detail of the experience, the places where the child may find themselves in a certain type of situation, and areas where that child might have a particular feeling or a response to a situation. For example, your social story might include the sentence 'at the park I will need to wait in line to use the slide. This may feel frustrating.' Social stories also emphasize rules, and policies of a particular place. In the sample social story below, you will see that a child must take off their shoes to access the gym. They are provided a point of control where they can choose whether to wear socks or go barefoot, but the strict policy of the gym is that no shoes are allowed on the equipment. The reason to include this in the social story is so the child is aware of this rule ahead of time, and if this is a behavior that needs to be practiced at home it can be reviewed there before entering the setting. It is important to end a social story on a positive or acknowledging note. This could be that the child tried something new, or that they had a positive experience at that particular activity or program. It is also important to include things that can feel reassuring for a child throughout the social story. For example, if their parent or a peer is allowed in a certain space to remind them that they will have access to that particular person. Also, to share if there is a bathroom or water, or snacks available, can be useful to help reduce certain anxieties.

How Can I Create a Social Story?

Social stories are easily created in Microsoft Word or PowerPoint programs. There are paid social story programs and free options on the Internet for templates and stories that are easy to design. If you are creating your own social story, it is important to have the background be as distraction-free as possible. The text should be in a large simple font, and without too many sentences on each page. Some people use clipart or other graphics for the visual cues. However, it is better practice to use pictures from the actual site of the activity or event in order to make it more recognizable when that child or individual arrives.

*Sample Social Story

The Children's Gym

Social story created in collaboration
with your local autism center

I am going to The Children's Gym!
This is a place I can play!

I can bring snacks and a drink from home.

When I arrive at the gym I need to take off my shoes. I can be barefoot or wear my socks.

At the gym I can choose to swing, jump, run, or climb!

There are adults at the gym help me learn how to use the equipment.

My family member can go on the trampoline with me!

There will be other children playing at the gym, and when it is noisy I can go to the quiet space.

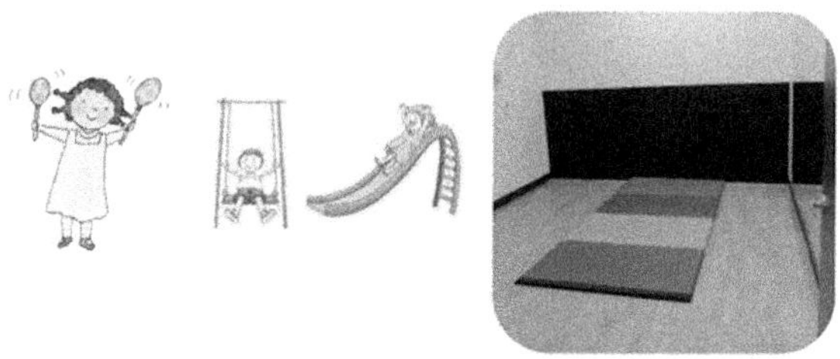

I might need to be patient and wait my turn to use the equipment.

When I want to take a break there is a table where I can color or play with toys.

The gym also has a bathroom and a water fountain that I can use.

I had fun at the gym and I am proud of myself because I tried new things!

Chapter 12

Designing Large-Scale Events

Businesses often host or participate in events. These may be a typical part of the organization's offerings, or they may be for celebrations, holidays, or special occasions. The scale can range from a small arts and craft fair of just a few hundred people to holiday gathers or even large music festivals where upwards of 100,000 people attend. While planning for these events, most organizations will automatically consider the needs of people of physical disabilities by incorporating ramps, large-size port-a-potties, and preferred parking into the event, but they do not often think about the needs of a neurodiverse individuals and their families. There are many aspects to consider when designing events to make them more accessible to people with autism, and these are the key areas to focus on:

- Location: The site choice is a significant decision. This will determine what is needed for safety, way-finding, and environmental modifications. Most importantly, site location will determine significant safety aspects that will need to be addressed, such as water, roads, heat, sun, structures,

DOI: 10.4324/9781003185406-16

and noise. Site also contributes to congestion, distance to transportation, ability to place a quiet area, and ease of communication with staff, vendors, and attendees.

- Safety: Water hazards are one of the most important safety concerns for children with autism at an event. Be sure to erect barriers that prevent attendees from going into water. Ensuring that play areas are far away from vehicles and moving rides is also necessary. A child with autism may not abide by signs that indicate not to go into certain areas, and if they are potentially unsafe, physical barriers should be installed.
- Staff knowledge and communication: Consider training volunteers with an autism overview. This can help them frame their job duties while considering the needs of all attendees. Provide multiple methods of communication to staff so they may best interact successfully with attendees.
- Marketing: When marketing an event utilize accessibility visuals and icons. This can help attendees know what supports are available for accommodations and safety. Including other visual supports or a social story when sharing event information can also be useful.
- Way-Finding and maps: When designing way-finding supports for an event, either through maps or with apps and digital technology, be sure to utilize visuals as part of the map. Visuals include symbols, icons, and colors that may indicate where a low sensory space is located, as well as identifying potential safety hazards.
- Signage: Most events will have extensive signage to help attendees know where to locate things such as concessions or bathrooms. Consider making signs highly visual with minimal language. Also, having continuity with colors can be considered a support. For example, the sign indicating the location of the bathroom is green and it is also green on the map. Use icons to identify quiet or low sensory areas.

- Quiet or low-sensory space: Quiet spaces at events can increase accessibility because an individual with autism may not be willing to attend if there is not a place designated or available to take a break away from the crowd and the stimulation. A quiet space should be located in an easily reached area, but be at a far enough distance from the crowds to offer some space and quiet. Quiet spaces should include seating, shade, and if possible water and a restroom nearby.
- Lost-child area: Large events provide opportunities for children and parents to become separated. Indicating a specific safe place where children can either go, should they become lost, is useful. This area should be staffed by workers who have at least some basic autism training and signage should be clear at the location and on maps of the event. Identify safe places, such as a guest services booth or tent, as a place a person who is lost can go to receive help and support.

Events in unique locations or that have potentially challenging aspects will require you to develop event-specific supports. For example, a new-product launch may include standing in lines near busy roadways or a process for sorting who is allowed to purchase certain items based on arrival and availability. Supports would need to include high levels of communication supports for directions, understanding timelines, and process, as well as safety considerations for adults who may be excited and less focused on spacial safety. For all events, include input from your Inclusion Team and be prepared to develop unique supports.

Chapter 13

An Example: Libraries

Most neighborhoods have libraries in their community. In some regions, this may be a small library, or a large library that is part of a county-wide library system that has extensive oversight. Because libraries are in most regions, and they serve people of all ages, they are a good type of organization to utilize as a sample for Autism Friendly strategies. The section will review physical space in the library, opportunities for sensory friendly browsing, and Autism Friendly events. Many libraries I've worked with do have a special needs or inclusion specialist indicated as part of their staff. Typically, this is a librarian or children's librarian that has received some extra training or has a degree in this area. Other libraries that are part of larger systems may have a special needs, disability, or inclusion committee or work group in place where librarians and staff from across the region will gather to develop programs and discuss the needs of their readers.

Sensory Library Spaces

Libraries tend to be bright primary colored bulletin board-filled spaces that are there to encourage, especially children, to

DOI: 10.4324/9781003185406-17

engage with reading. While this can create a fun and stimulating environment, that can also be a challenging experience for a child with autism. One way to help make your library more accessible is to offer a low-sensory space. The simplest description of a low-sensory space in a library setting is a room or a quiet corner where a patron or reader is able to take a break. Please recognize that a low sensory space is different than a sensory space. A sensory space is designed to engage the senses, and a low sensory space is designed for the senses to be encouraged to be calm. For libraries that have the ability to dedicate a specific room for this use, that is the best option. It may be possible that a multipurpose space can have some sensory friendly design elements incorporated into it. However, it is important that that space is accessible.

Since we discussed some general requirements for sensory friendly spaces earlier in this section, ideas that have been used in library settings will be shared. It is important to identify the room or the area as a sensory friendly zone. Signage should include some visuals and provide some basic directions for the space using those visual supports previously described. The physical space itself might include things like beanbag chairs or soft comfortable sofas to sit on. It should also include a lighting design that allows for flexibility, including the ability to dim the lights and not utilize harsh overhead lighting. It is important to consider ambient noise in the space as libraries sometimes have computers and air-conditioning systems that might hum or make other sounds. The space need not be too large, as you don't want it to fill up with multiple people. It should be approximately the size to accommodate one family. Options for fidgets or other hand-held items to help individuals express their feelings of overstimulation would be useful in such a space. If the library has the budget, offering noise-canceling headphones is a nice option. It may also be helpful to have a stack of blankets or a mat on the floor in case a child would like to lay down for a few moments.

Some libraries require that a family sign up for a time slot in the space; however, this has the potential to limit the accessibility when someone is having an unscheduled moment and just needs five or ten minutes in the space. Libraries might incorporate the option for soothing music, and tactile items on the walls. It is important, however, to make sure that the space is not a high sensory experience. We don't want to fill the walls with lots of colorful items that defeat the purpose of a low sensory option. Consider a child's safety in such a space and be sure that anything that is accessible is soft and that things like outlet covers are in place and furniture does not have any sharp corners.

Sensory Friendly Browsing

One method of improving access to all patrons and readers at your library is to offer sensory friendly browsing hours. These hours take place outside of the hours for the general public. This strategy is often utilized as a way to encourage individuals and families to come into the library. Too frequently a library will put these hours at one particular time of day which can be quite limiting for a family. For example, some libraries that open at 9 AM in the morning will have 8 AM to 9 AM be their time for sensory friendly browsing. However, this may be a time of day were a child is not yet awake, or in the middle of a breakfast or morning routine. It is important when designing these hours and your schedule, that there are several options. It may be that one day a week is early in the morning, but there are other times in the evenings or on weekends. This will allow more people to participate.

Autism Friendly Programming

There are many libraries that are offering Autism Friendly programming. While we are going to discuss some options for these programs, please remember what was mentioned previously in this book that inclusion means everyone, and while these programs can certainly be beneficial, they are a type of segregating a specific population from others. Best practice is to incorporate accommodations and supports into programming that is offered to everyone. The suggestions made here can be utilized for both types of programs. A sample registration form is provided and can be modified depending on the type of program.

*Sample Autism Friendly Program Registration Form

INFORMATION FORM

Child's Name: ____________ Parent Name: ____________

What is your child's diagnosis or area of support needs?

____

What is your child's learning style (circle one)?

visual auditory experiential other ______________

What is your child's primary method of communication?

____

Do you have concerns for your child's safety? (please let us know how we can support you

____

What is your child's most comfortable learning setting? (circle one)

one-on-one learning	small group learning	large group learning

What are your child's strengths, interests, or hobbies?

____

Does your child have any behavioral concerns that you would like to make us aware of?

____

What areas do you suspect your child will require some assistance with?

□ Adhering to the schedule	□ Understanding the assignment	□ Attention
□ Following Directions	□ Social Interaction	□ Motor Skills

Is there any information that you would like to share to help us know your child better?

____

Will you or another caregiver be present for the program? If so, please indicate who the caregiver will be during the program.

____

Sensory Friendly Movie Night

Libraries will often have events where they pair a movie and a book together to offer a movie night. Introduce the book and provide different reading and literacy level options for participants. These can range from picture books to graphic novels to complete works to audiobooks. This way your participants may participate on the book side of the event regardless of their reading level or ability. When showing the movie, some suggestions are to have some lights on in the theatre space, to have the volume of the film at a slightly quieter level, and to provide many options for seating that go beyond folding chairs.

One way to incorporate some autism-specific strategies would be to design a social story about going to see the movie, and to also create a heads-up type playbill that shares where loud noises or unexpected scenes may happen in the movie. It is also helpful to remind families of where the low sensory space or quiet room is in case an individual needs to take a break. Also set some ground rules for physicality during the movie. This could be that participants can stand up during any point in the movie but they are directed to not walk in front of the screen or projector, or that if they need to walk around during the film that there is a space provided toward the back or the side of the theatre or room. It is also an option to suggest to families that they bring their own pillows, blankets, and food, if food is allowed in the space. It is also suggested to consider potential developmental levels of the viewers, and when advertising the movie to be sure and target your audience by indicating not only the rating of the movie but if the movie should have particular language or violence.

Sensory Friendly Story Time

Sensory friendly story hour or story time is very popular in libraries. These times are typically during the weekdays for children that are home-schooled, or Saturday mornings. Sensory friendly story hour can be an excellent introduction to the library for children. It is an opportunity for them to be in a space with other kids that are their peers and to have social interaction while learning about books. This type of program should be developed with some goals in mind. Is the goal of the experience to simply bring people into the library setting, or is it for learning or developing a specific interest? Some sensory friendly story hours are higher sensory. They utilize the experience in a book and then amplify it to be highly interactive. An example of this is utilizing a bubble machine while reading a book about sea creatures. Perhaps also passing around shells that have different textures, and maybe utilizing stuffed animals or puppets that are the type of fish in the book. These types of experiences can be very engaging. A lower sensory story hour might be one where the lights are dim, the children sit on mats or blankets on the floor, and a more soothing type of story is read slowly and at a lower volume. It is possible to distinguish between these types of settings when marketing this type of event.

All sensory friendly events should have options for a quiet or break space and for plenty of room for the child to move around. These types of events also have the opportunity to show children and families the children's section of the library and to possibly encourage kids to become a library card holder and possibly check out a book before or after the program. If library staff feel like they need support in developing some of these programs, especially if there is a particular curricular outcome goals, it can be helpful to engage autism specialist, specifically those that have occupational or physical therapy expertise for our speech-language pathologist.

Chapter 14

Summer Camp Inclusion/IRPs

Summer camp can be a life-changing experience for children. It offers them the opportunity to be independent of their parents, and to connect with peers their own age, while being outdoors and doing activities that are not available in a school, classroom, or home setting. Often the friendships and affinity for camp programs are lifelong. As a camper alumnus myself, my experiences as a camper are some of the most important of my life. Often when the families I've worked with have children that attend camp, they are different after the experience and have more independence and a broadened scope of interests.

When designing an inclusive summer camp experience creating specialized staff positions is essential. Many camps now have a full-time inclusion specialist who focuses on diversity and disability. If your program is unable to fund such a position, then it is essential to have at least one point-person or someone at the director level who has this kind of training. Ideally, this individual will have a background in exceptional student education, or have a behavioral specialty.

DOI: 10.4324/9781003185406-18

All staff that are working with children in the camp setting should be educated about neurodiverse campers. An overview of autism is essential for everyone as they are likely to work with these children in at least some capacity. For counselors and program staff that will be engaging closely with children, having more in-depth training is important. This training should focus on communication, behavior, social skills, general safety as well as water safety, elopement, interventions, sensory concerns, and overall support techniques.

For program managers who feel overwhelmed by the prospect of full inclusion, there are other ways to support children in your camp without necessarily taking on the extent of the education that might be needed. An alternate option is to bring in experts that have experience working with children that have high support needs and have them be those children's point people. This is a way to integrate professionals with a high level of experience in this area, and be able to offer some form of inclusion. For example, in my region, there is an organization has a program where they are able to provide highly trained counselors to be one-on-one supports for autistic children in a summer camp setting. This option allows the child to participate without the camp itself taking on the challenge of providing long-term training and intensive training to counselors they employ.

When enrolling children into camp, the standard camp form should ask a few questions about the needs of the child. The answer to those questions then trigger a more in-depth review for children with autism that are going to have higher support needs. The second step of the process is for the child and family to meet with the camp director or specialist to design a safe and supportive environment. At the City of Boynton Beach, Stephanie Soplop, who was the recreation program specialist, and is now the ADA coordinator, created a process to help meet all objectives. She developed something called an IRP. This stands for individual recreation plan. An individual recreation plan is modeled after an Individualized Education Program (IEP). A sample IRP is below, and it identifies the child, their emergency contact information, their interests and likes, any challenges and limitations they may have, safety concerns, and it outlines strategies. The strategies are in direct response to the support needs of the camper. It leaves rooms for comment and for parent and guardian signature as well as a signature by the camp.

Other aspects of an inclusionary summer camp program are to have frequent observations of how the child is doing, and to adjust the IRP based on those observations. It is possible that the assumed needs at the beginning of the program will adjust overtime as that child becomes comfortable in the setting. Visual supports, social stories, and other strategies are also useful to help the camper integrate into the camp programs, and can be developed for specific aspects of the experience. For example, a social story about the first day of camp would be useful, but also creating social stories perhaps about the first time on a boat, or a visual support on how to make s 'mores over a campfire, are helpful.

*IRP (Individualized Recreational Program) Sample:

City of Boynton Beach
Recreation & Parks Department

Individualized Recreational Program (IRP) Plan

Participant's Name: Participant's Age:

Parent/Guardian Name: Home Phone:
Work Phone: Cell Phone:

Interests/Likes: Being a helper, playing games, playing outside, sports

Challenges/Limitations: Difficulties staying still and following directions.

Strategies

Assign "jobs" so he can be a helper such as holding the door, sweeping the floor, helping to carry equipment

Use behavior chart with stickers as a reward and give him reminders of the camp rules

Allow him time to sit and color in his coloring books he brings from home

Point to Stop Signs displayed on the doors as a visual reminder not to leave a room without a counselor
Remind him of the boundaries, to stay with the group, and what area he needs to stay in when outside.

Use prompts prior to transitions to different activities, being clear and specific
Examples:
In ten minutes we will be lining up at the front of the room
In five minutes I will be calling everyone at this table to line up

Always keep an eye on him and give him praise when he is following directions.

Comments:

____

____

____

Parent/Guardian ______________________________
Print Name Signature Date

Recreation Supervisor ______________________________
Print Name Signature Date

Chapter 15

Employment and HR Tips

While this book primarily focuses on the needs of customers interacting with a business, there are definitely going to be people with autism that are employees, or that want to work for your business. The next book in this series is Autism Friendly Employers: Diversify Your Workforce, which will include more in-depth strategies about working with employees with autism. However, it is useful to briefly touch upon this here and for human resource departments to consider accommodations and modifications to their typical processes in order to improve and develop inclusionary practice.

Accommodations

One of the most simple and straightforward ways to make changes and to ensure you are well informed about employee accommodations is to utilize the United States Department of Labor's Office of Disability Employment Policy (ODEP) Job Accommodation Network at https://askjan.org/. JAN provides

DOI: 10.4324/9781003185406-19

excellent information and even a helpline to assist employers on providing accommodations. Accommodations to support the employee in completing essential functions of the job they may suggest include:

- Equipment
- Environmental modifications
- Flexible scheduling
- Flexible job tasks
- Other unique solutions

Hiring Process

Other ways to improve accessibility is to be open a to modifying some of your processes, especially around hiring. Job interviews are a very social and verbal activity. For people with autism, these can be places where prospective employees experience deficits. Consider some very simple adjustments to your interview process. One modification I recommend is to provide the interview questions to the individual ahead of time. We may think that the answers to interview are supposed to be something that we come up with answers on the spot for, the majority of the questions we ask including things like what are our goals in five years and why do we want this particular position, are concepts that are important people have time to put some thought into. It is likely that you will receive a more thoroughly developed response if the applicant has an opportunity to organize their thoughts. Another suggestion is to have the questions printed out and available to be in front of the person during the interview. This can be helpful because it is possible that the applicant is experiencing stimuli from the environment that is competing with the voice of the interviewer. Another area to be open and flexible is that applicants may not have a traditional résumé. A resume may be a functional resume, which highlights skills and abilities versus work history, or it could be a video résumé or portfolio that shows the applicant completing a job task, versus describing those tasks. Non-traditional resumes are very commonly used in the disability community and should be received with the same amount of respect and consideration of a traditional résumé.

Vocational Rehabilitation

If you are in the United States, there is a national program called Vocational Rehabilitation. This organization helps people with disabilities obtain and maintain employment. They can provide job coaching and support for accommodations needed in the workplace. If you are having difficulty meeting the accommodation needs of an employee, for example, they need an adjustable desk and it is not in your budget, it is possible that they are a VR customer and that they are eligible to have an accommodation funded. There are many resources that are available to support people with disabilities in the workplace, and it's important to have a good understanding of what exists and can be accessed in your community.

Chapter 16

Conclusion: Share with Others

Start small. Celebrate wins. Observe those you are impacting and take the time to see the significance of this work. If those moments touch your heart and light a fire in you, you will understand those who are making this movement the mission of their careers and their lives. Listen to the experiences of businesses that created spaces such as barrier-free parks, art in public places, and who hire interns or develop mentoring programs. Expand these ideas from those with autism to the larger disability community. Develop policies and procedures that support the continuance of this work in case you leave your position or employer. Create community work groups and conferences that highlight your efforts and gift your strategies to others. Tell your story while it grows.

Success Stories

Success stories are evidence of impact and show proof of concept. They are demonstrations of how an idea works after it is implemented, and how it improved or influenced an

DOI: 10.4324/9781003185406-20

experience. Success stories are an excellent way of highlighting an individual or group's contribution to the community. Some success stories might be focused on a specific child or family. In those success stories, it might be that the business provided something specific that now allowed that child to have a better quality of life. For example, if a dance program or craft program or swimming program became inclusive, where in the past they might not have been able to accommodate a child with autism, it could be viewed as a success story for both the business and the family.

Other success stories might include when an individual with autism gains employment at a business. Perhaps that person has always dreamed of being a camp counselor, or a professional in the administrative department, or a first responder. The business has learned about inclusive practice, they hired that individual, and perhaps are providing some supports on the job, such as things like mentoring. Other types of successes might include reporting out on events that are disability-specific or highly inclusive, such as people who develop a new program or when a social story or specialized visual is created. Success can come from inside the business walls, or from those that reside in your town.

Conference/Expo

A few years ago I started to develop and host the world's only known Autism Friendly Institute at FAU's Center for Autism and Related Disabilities. At this conference, business and city leaders are able to share the ways in which they have implemented strategies to make their community more inclusive. The institute started small, but even in early days, attendees participated from outside our region and left with ideas and concepts they implemented. The participants in these institutes were employees, management, and company leaders. The conference has varied from year to year and typically includes one keynote speaker, a panel of representatives from business or community projects, and a brief update by me sharing general strategies and new ideas that have come into light over the past year.

Keynote presenters have included the City of Boynton Beach, a police sergeant from the City of Jupiter who was sharing information about their 911 registry program, as well as Dr. Toby Honsberger who was the coordinator of our local Autism Friendly nature area program. It is useful to have a panel as well in order to bring more voices to the table. Panels have ranged from three or four to six professionals who are doing something new in the community or who have a a unique neurodiverse perspective. It is useful for these panels to include a person with autism who has been directly impacted by either a strategy or program. It is helpful to highlight how it has impacted their life and why it has made a difference for them.

These conferences can be offered with minimal expense, or can be sponsored by local organizations. Highlight individuals who were implementing the strategies and that are willing to share their stories without charging speaker costs. Offering them a marketing opportunity or exchange will be a way to support their needs. During Covid times, our institute moved online and while it was a different feel than an in-person experience, the number of people that were able to participate, and the expanded distance from which they were able to engage, definitely magnified our reach.

Declaring You Are an Autism Friendly Business

There are many ways to make this declaration, to tell the world you are now Autism Friendly or a Neurodiverse Business. Many businesses will do this with a press release, a newspaper article, or media story. Other ways are to introduce it at a commission or leadership meeting. There are people out there who will certify you. I've always stayed away from this term as I feel like it is more of a designation, than a certification. As a designation, it is important that the process continues to change and grow overtime, as the needs of your customers and business change. They are ways to create materials that identify you with this designation. Some businesses will develop a logo, or graphic, that indicates they are autism or disability friendly. Others will create a website that is specifically for Autism Friendly programs, which highlights the changes they made, strategies they implemented, success stories, and their plans for the future. All marketing options are a useful way to engage customers and the community in a dialogue about neurodiversity with your business.

You might wonder if you've done enough to declare that you are Autism Friendly. It is important to acknowledge and celebrate what you've accomplished, as long as you realize that as the community changes, there may be more to do. The reason for having your inclusion consulting team is to help inform you of shifts in the community and new strategies to implement and consider.

Tell the World

There are many creative and unique ways to tell others that you are Autism Friendly. For example, a fun way to share that your business is Autism Friendly is to wrap your vehicles in information and graphics that highlight this commitment. Many businesses will include things on their bumpers or the side car doors of their police cars, or on food trucks, business cars, or emergency vehicles. Some businesses will develop an Autism Friendly graphic or logo that that will incorporate into their website, email signatures, social media, and on marketing materials. Consider sending out a press release to your local media and possibly having an event or ribbon cutting showcasing your Inclusion Team. There are businesses that will create a short video explaining how and why they are now Autism Friendly. I've even seen organizations create t-shirts that their team wears during certain times of the year. The possibilities are endless.

Appendix: Examples

There are many varieties of businesses in existence. It would be impossible to offer samples for every kind of company, but I am going to provide suggestions for the types of organizations that at least fall into the main categories of businesses in the community.

The following examples are based on real-world experiences. Different businesses and cities created these events, products, and initiatives, and while they may personalize them to the needs of their own community, they are models of how to implement strategies for inclusion. Samples are provided within the examples, as available. Please modify them to fit your needs.

Retail

Retail businesses are all about space. Especially when those businesses are physically in person versus online. For a customer, those environments are sometimes challenging to navigate especially as some businesses use flashy designs and bright colors to draw people in. The first area retailers can consider is how to utilize visual supports. This can be done by having different categories of items be identified utilizing shelf colors, background colors or designs, and also visuals that are representative of the words. For example, if you have a retail store that sells soap and body products, having the different types of soap identified on shelves that are colored can help someone navigate towards that specific category of what they need. For example, having foaming soap on a green shelf, and gel soap on a blue shelf, and separating things out like soap, shampoo, and conditioner, in different areas or having different smells and flavors of products in spaces together that are also identified either by colors or design or visual supports. For example, if your shampoo smelled like vanilla, not just having the word vanilla either on the product or on the label, but also having a picture of a vanilla bean or something that represents that flavor lemon. You can then use this visual across all your products that are lemon-scented.

Sounds and lighting can also be challenging in a retail space. Almost rear tellers will use bright overhead lights, there are some that have the opportunity to create a cozy setting with lamp lighting or lights that are not so bright white but perhaps more yellow or blue. Sounds can also be distracting and create sensorily overload in these types of places, and it is important to consider how loud your background music and air conditioning vent systems are in relation to what kind of space you were offering and how much that sound will bounce. You might look at adding carpet or other sound-deadening options to help reduce the extra noise in your space. You can also be welcoming to people trying to navigate the senses by indicating on small signs near your front door that wearing headphones as welcome, or can be even encouraged. One of my personal favorite things to do is to wear noise-canceling headphones with fun music while I grocery shop, as a completely changed experience.

Other visual supports that retailers can incorporate are around the checking out and payment process. It can be useful to have simple directions that are offered in pictures and icons and not in words near the register. For example, signs that indicate how to pay by credit card and arrows that show where to stand on the lines get long. Also. if you offer self-checkout, providing at least simple visual directions can be very useful.

Retailers can also create sensory friendly browsing hours, especially when they have store spaces that are particularly high sensory. During these times lights are dimmed, music is quieted, extra support can be provided, and space and time to browse at one's leisure can help to create a welcoming experience.

Restaurant

See chapter on visual supports to create icons and visual supports on menus. Other options to be more inclusive is to offer table choices. For example, a table near the front door or near the kitchen will be a high-traffic area, while tables in the middle of the room may have an increased noise level. Some restaurants identify low sensory tables which may have reduced lighting, be away from sections where patrons or servers frequently walk back and forth, and are far from a noisy kitchen.

Hotel

There are many hotels that have taken on being Autism Friendly. One of the first had been featured on their local evening news, and I showed the clip as part of my training. Hotels can implement strategies by offering Autism Friendly rooms. The rooms may offer any or all of these recommendations. First consider the bed options. Can the beds you provide have safety rails attached? Is it possible to offer a roll-away bed that is low to the ground as an alternative to a high bed? When considering safety, installing door alarms can be extremely helpful. Door alarms are inexpensive and can often be toggled to be turned on or off. Door alarms should be installed on the door leading to the hallway and on doors leading to bathrooms. Another safety concern is to make sure that breakable decorative objects are bolted or glued down. Television sets are mostly mounted on walls or to furniture, but if your hotel has free-standing TVs, consider securely attaching them to a stable surface.

Personal Trainer

It may be difficult to consider how a personal trainer could be more Autism Friendly, but even these types of small business services can improve accessibility. Before an initial training session, there may be some belief that the client will know what to expect. There are actually many variables to services like this, and some or all may be foreign to a new client, which may be a source of anxiety. Areas to consider are equipment, space, environment, clothing, duration, and activity. For equipment, try and provide a specific list of what will be needed. If it is possible to send links to a place where these items can be purchased, that can be helpful. Also, if no equipment is needed, be sure to share that. Space can be thought of as the physical room or area the workout will occur in. It may be that the room has furniture that needs to be moved or even is a central place in a home where there will be visual or auditory distractions. Be sure to be clear about the duration of the session and provide information about options to end early if needed. Discuss safety concerns which may include hydration, and if the session is outdoors, possibly sunscreen or protective clothing. It is helpful to be very specific about the type of space that will be required if the individual needs to create that environment in their own home. Also, discuss any safety concerns, fears the individual may have, and the best ways to communicate throughout the experience.

911/999 Emergency Registry

When an individual makes a call for emergency assistance basic information such as location will typically register on the computer of the receiving operator. While this is useful for aiding in locating someone, it provides no other helpful information. If the person is calling emergency in the region they live, a 911/999 registry can support a better experience for both the family requesting help, and the first responders that arrive on the scene. While registries are completely voluntary, they are frequently appreciated and sought after, especially by parents of children with autism that frequently elope.

Some communities may already have a version of this type of registry that they utilize for older adults that have dementia or other cognitive concerns. For some regions, it might be easiest to modify and adapt their current templates and use them for this purpose. For cities that do not have a registry system, it can be something relatively easy to create and incorporate into the address and location features of your emergency response system.

The important information to gather is the individual's name, identifying features, their age, their primary method of communication, what may frighten or scare them if approached, and the specific location they tend to go to when they elope or leave the house unsupervised. These locations may include places like the neighborhood playground, pools or other bodies of water, friends' homes, or other places they are familiar with. A child may not be cautious about crossing the road or may enter the water without knowing how to swim or be safe. Having information ahead of time about where a child is likely to elope can potentially save a life.

A useful piece of information to identify is the individual's primary method of communication and to know whether or not they are verbal or nonverbal, and if they have basic literacy skills. If your emergency system allows for photographs, that can also be very helpful. It is necessary that families update this information on a regular basis, and it is suggested that this is done annually. When a family calls emergency response, and the first responder in the field can bring up all of these details, the odds of locating a child or adult more quickly increase.

An Artist Illustrates the City Calendar

As the Autism Friendly Cities program in Boynton Beach expanded, they were interested in connecting with an individual with autism who could bring their talents to a city project. When they started brainstorming ideas, they were unsure of how to locate such a person. They turned to their Inclusion Team for suggestions, and discovered a Familial Entrepreneurship called Artists with Autism (www.artistswithautism.com/). The city reached out and connected with artist Brandon Drucker. After discussions with him, it was agreed that Brandon would use his artistic abilities to illustrate the annual calendar published by the city. Not only would Brandon illustrate the cover, but he would provide art for every month of the calendar. To make the calendar more locally focused, Brandon created images of the city's intracoastal park, Italian ice and creamery shop, and the marina. The calendar was an extraordinary success, and the city published and distributed a massive 75,000 copies. The calendar was entered into to a nationwide competition where it won second place out of over 700 submissions. Brandon was recognized at a public meeting of the city commission. At that gathering, a woman who creates quilts saw the calendar and offered to print the art onto fabric and sew it into a quilt. The quilt had the panels from each of the different months of the calendar and after it was acknowledged by the city, the quilt hung in the city of Boynton Beach's main atrium that year. The city calendar, illustrated by Brandon, is a powerful example of how to highlight the talents of a person with autism who lives in your city, in a creative collaboration, which supports your city's inclusion commitment, as well as supports a local artist and shares their gifts and their name with the community.

Autism Friendly Nature Area

Natural areas, forest preserves, and conservation parks, or wildlife sanctuaries are frequently part of community spaces and fall under the auspices of a park and recreation or public works department. These spaces are beautiful and inviting, but also have many safety concerns for individuals with autism. The most troubling is water. According to the Autism Society of Florida, *children with autism are 160 times more likely to experience nonfatal and fatal drowning than their neurotypical peers, and in the State of Florida, the number one cause of death for people with autism is drowning* (www.autismfl.org/drowning-prevention). There are many ways to enhance safety measures throughout these natural areas without compromising the environmental appeal. Ways to encourage engagement in the natural area can also be incorporated to improve inclusion.

In South Florida, a natural area was designed to be Autism Friendly by Dr. Toby Honsberger at The Learning Academy at the Els Center of Excellence in collaboration with Palm Beach County Department of Environmental Resource Management and the North Jupiter Flatwoods. The experience begins on the website (http://tlacad.org/flatwoods/) where visitors can explore the nature area through an interactive virtual tour. This is extremely valuable for individuals who might want to know what a space looks like before they choose to visit. A second virtual experience is the social story or narrative that they developed. This brings participants through the park section by section with pictures and first-person text that describes what they will see and experience. This also includes some basic rules and suggestions for navigating areas that may have safety concerns. A second version of the social story is offered that includes symbols for those that have limited reading abilities. There is a section on the website where trail rules are clearly indicated, and they also developed a nature trail scavenger hunt and sensory scavenger hunt.

In the nature area, there are paved trails that provide boundaries for walking and enjoying the space. There are visual supports incorporated into their signage. There are frequent benches that allow for resting throughout the nature area. They've also incorporated a swing and some sensory stations to encourage a real engagement with the space. Barriers were installed near water areas and other places that may have safety concerns, and a map is provided that is clear about where each path leads, the distance of each path, where the rest areas are, and what is available to interact with on that path. The thought put into this experience was extensive, and this space is utilized by adults, children, families, and local schools.

Inclusive/Barrier-Free Playground

Playgrounds are designed to be enjoyed by children. A child should not have the experience of going to park and not being able to play. However, many recreation areas contain barriers for individuals with autism and they are unable to fully participate.

Barrier-free park is a place for children to play, people together, and adults to enjoy the kid at heart that's inside of them. Many parks have barriers that prevent accessibility for people with autism and disabilities to participate in accessing them. The park may not be conducive to everyone utilizing the equipment or being in the type of space.

There are many ways to remove those barriers and increase accessibility. These are the primary areas that need to be considered when developing this type of space. Safety is most important. If the park is near open water, or a swamp, or river, it is essential to develop measures that will make sure a person with autism cannot access the water in an unsafe way. This may be done with fencing, strategically placed bushes, signage, and changes to the texture of a path to indicate that something different is coming. Also, any social stories or visuals that are made related to the park need to identify any potential safety concerns that a parent or caregiver might need to be aware of. Another aspect to consider is the open areas of the playground. In many playgrounds, all of the activities are bunched together, which creates lots of stimuli if there are several children using the same piece of equipment. One way to help mitigate this is to create areas that are slightly separated to provide more physical space and opportunities to be together in the same general park, but to have less physical interaction that is so close to each other. Playground equipment itself can become more accessible when things like height and weight restrictions are increased. There may be a teenager with autism that really likes to swing, but if the swing is designed for a five-year-old child, that then becomes something that is no longer accessible to them. It is important to recognize that individuals may be at different developmental stages, and that the facilities at the park should be accessible to all people.

Another aspect to think about is the textures of the ground in the play area. Is it going to be concrete or cement, or will there be natural grass or asphalt? These are all things to consider as some regions may need certain materials based on climate. However, recognize that the different types of materials that are used can create a sensory experience for someone. For example, if a path to a park is lined with small stones, the experience of stepping on those rocks and the sound of those rocks is going to potentially be impactful. Spaces can also be designed as visuals by using different types of materials to indicate where to play, where to sit and rest, and what pads are there to get in and out of the park. Other things to consider are with all of sidewalks, paths, and bridges. It's also important to note that bathrooms can be beyond the typical ADA requirements, and enhanced with things like toilet seats that turn, and large stalls to allow for a caretaker to be in the space with an individual.

Having a quiet space that is lower sensory in this environment is also useful. It might be slightly away from the main playground area, and would be best to be in a type of pavilion that provide some shade and coverage from weather, as well as a comfortable place to sit, should an individual or family need a few minutes away from the action.

Autism Day of Fun

One way to support the customers in your community with autism is to create an Autism Fun Day. These types of events are titled differently depending on where you are located, and they can range from a family fun day, to a day for autism, to an autism walk, to awesome for autism day. People get very creative in naming their events.

An Autism Day of Fun event typically includes your local autism-related agencies, support organizations, and service providers. Sometimes these events are developed to provide some fundraising for these groups. The event may include things such as music, the firefighting trucks, and equipment, bringing police cruisers out for children to see, and possibly even the police dogs. Activities might include arts and crafts, games, a bounce house, or even a chance to meet a local celebrity or famous person who visits with the families.

If your event includes a walk or 5K, it is helpful to create a couple of different options for individuals and families. Sometimes there will be a traditional 5K and a secondary walk, of perhaps one lap around a park or a shorter designated area. When determining time of year for this type of event be sure to consider the weather, and how that might impact having people outside for long periods of time.

Typically, the agencies and groups involved will have some sort of resource aspect to this type of day and will often offer giveaways or provide activities. When designing these types of events, it is important to ask your local experts to provide some of the activities as they may be doing this as part of their practices and it would be useful to have activities designed by people with a strong background in autism to offer at this type of event. This is also a chance to integrate your local social enterprises and family entrepreneurships, so that they can perhaps sell some of their goods and be included at this type of community event. This is another time where safety and utilizing event planning strategies for inclusive programs are important. Sometimes we put events into a park or natural area and don't recognize that we have just asked 100 children with autism to play 20 feet from a pond. You need to make sure that you pick spaces responsibly, and that you have safety as our number one concern.

These opportunities are also useful for businesses that might be Autism Friendly to showcase some of the ways that they are. For example, in my region, there is a business that developed a sensory friendly bounce house. They spent many years creating this inflatable, and received input from local autism specialists and families. Sometimes these types of events are incorporated into holiday activities. For example, some businesses offer a sensory friendly visit with Santa, a Thanksgiving meal, or an Easter egg hunt. These are ways to connect with families during special times of the year and to ensure that celebrations are not exclusionary.

These events also provide opportunities for families with similar experiences and needs to gather. Bringing them together with some sort of structure is often overlooked. Incorporating some sort of family meet and greet or opportunity for children around the same ages to socialize and be together can be extremely valuable for these families.

Outreach and Giveaways

Everyone enjoys getting something for free. While that is nice, it is helpful when a giveaway is also useful. It is a win-win when it is something valuable to both the giver and the receiver. When connecting with customers and families in your local community, there are things that can be given from the business to those customers that is helpful. Most of these types of giveaways revolve around safety, while some things can easily be purchased online or developed in-house. They can be given away in groups, sets, or kits, or as individual items that are offered at public events. Some items may be things that police officers and other first responders have in their possession for times when they connect with families and or individuals with autism.

There are many more items than those that are being suggested here, however, this should give you a place to start, and a way to brainstorm what may be needed that is more specific to your local community.

- Resource Packet: information about swim lessons, mental health resources, emergency phone numbers, and your community's crisis line. This packet of literature can be printed in full, or created in a digital format to save paper and costs, with than a website or QR code on one printed card to share with customers.
- Magnets: Emergency numbers for the local crisis line or places like poison control can be created and shared.
- Fire Decals: Companies make glow-in-the-dark fire stickers which are typically placed on or near the front door of a home as well as near the bedroom door of the individual

with autism. They say that a person with autism lives in the home, and they are helpful when firefighters are needed at a residence.

- Seatbelt ID Card Holder: Several companies, and creative sellers on sites like Etsy, sell small card holders that attach to a seatbelt. These are useful for first responders in a car accident or other vehicle-related response. A person with autism can have a card on their belt indicating they have autism, and sharing their name, age, communication preferences and abilities, and any medical needs that the person may not be able to express during an emergency.
- Wallet Card Information: The wallet card is a program through The Disability Independence Group, designed in conjunction with the Coral Gables Police and University of Miami Nova Southeastern Center for Autism and Related Disabilities. The purpose of a wallet card is for an individual with autism, a child or adult, to have a card that they carry on their person that indicates their name, disability, age, verbal ability, contact information, and emergency contact numbers. They also have the ability to add customized information. The wallet card is at no cost and is a way to have an item that identifies that a person has autism, without putting the indicator on a driver's license or state ID. On the website there is a video explaining the usefulness of the wallet card: https://www.justdigit.org/wallet-cards/
- Pool Watcher Tags: These tags, created with a laminated card and lanyard to go around one's neck or body, are used to identify the responsible adult who is tasked with watching the people swimming in the pool during a group gathering at a home or community pool that does not have a lifeguard. These tags are extremely helpful especially if the gathering includes loud music, many people, and alcohol consumption. When drownings happen, adults often believe that there was someone who was supposed to be the pool watcher, but it is rarely clear who that

person was. These tags can eliminate confusion and prevent tragedy.

- Electric Wireless Door Alarms: Door alarms are an essential safety item for families that have a child with autism that tends to try and elope. These small, inexpensive, battery-operated door alarms can alert a parent or caregiver that an individual has successfully opened a door and has possibly exited. These are useful in homes and for families that travel. Utilizing a door alarm can prevent a person from being in danger.
- Stop Signs: Adhesive small stop signs can be used by families as visual supports to prevent a person from entering an area that is unsafe for them, or by opening a cabinet or door that leads to items that could injure a person.
- Safety Proofing Items: Safety proofing items are those that can prevent an injury in the home. We may call some of these child-proofing items, but it is possible that they will be needed in a residence where an individual with autism may be a teen or adult. items include corner guards for sharp furniture corners, safety locks for cabinets and closets that may contain poisonous items, door knob covers, appliance or stove locks, and guards or covers for electrical outlets. These items are often small and inexpensive. They can be purchased in just about any home goods or large box store. They can be given out at safety events, or part of a safety kit.
- Children's Books on Safety: It is a nice gesture to include children's books in giveaway boxes or bags. These books can range in topics, but there are many that are created specifically for children that are safety-focused. These books can include stories about pool safety, fire safety, and stranger danger. There are many organizations that give these books away for no cost, or they can be purchased in bulk for a reasonable amount. It may be useful

to find some that are location specific to your business. For example, in the Florida region offering a book about hurricane safety would be relatable and appropriate.

Emergency Shelters

There are a variety of situations that can cause a need for your community to have access to emergency shelters. Conditions might include a hurricane, earthquake, flood, significant power outage, or other natural disasters. Businesses may host spaces that are used for shelters, or may have the need to direct their customers or guests to shelters, this especially applies to the hospitality industry. Most importantly, shelter staff may be volunteers or public employees that could be working outside of their scope, and it will be essential to make sure they have training. At the very least, they need basic information on autism, as well as how people with autism may respond to an emergency, and simple strategies and techniques that they can implement in a shelter environment.

Materials can be supplied to shelters to improve the experience for individuals with autism and their families. Comfort items will include blankets, pillows, and stuffed animals. Shelters can be supplied with earplugs, headphones, and fidget items to help with overstimulation. If a gymnasium is being used as a shelter, gym mats can be used to create a cushioned area. However, that type of space may reverberate sound, and if classrooms or other smaller areas can be offered to a family, that may help them have more space and less stimuli. Sensory bags, games, and pre-charged electronic devices that don't require internet can help with engagement and distraction from a potentially stressful environment. Social stories and visual supports should also be created to provide families when they register to be included in a shelter list, or when they arrive at the location.

Support Social Enterprise and Familial Entrepreneurship

One manner of increasing connection to people with autism is to reach out to your local social enterprises. You may not realize, but frequently there are many businesses and programs that are either run by or have employees with autism and other disabilities. A good place to start looking for these types of businesses is to ask your inclusion consulting team who they are aware of and possibly work with. Some of these social enterprises will exist within schools, transition or independent living programs, or they may be independent businesses started by relatives of individuals with autism, known as Familial Entrepreneurships.

Some examples of social enterprises and Familial Entrepreneurship, local to South Florida, are ScentsAbility Candles and The Chocolate Spectrum. These two businesses were started by mothers in order to provide employment and social opportunities for their adult children who have developmental disabilities. The businesses are both thriving and now employ several individuals with disabilities, as well as provide vocational training opportunities and life skills. The Chocolate Spectrum (www.thechocolatespectrum.com/) is a gourmet chocolate company that can individualize and design chocolate treats for any occasion. ScentsAbility Candles (www.scentsability.org/) is a candle company that can create unique and beautiful candles which can be enjoyed at home or used as corporate or event gifts and giveaways. There are many entrepreneurship programs that exist within schools, and locally, that includes students who make organic dog treats, one that bakes cookies, and at another school they have a woodworking shop where every holiday season they create large reindeer that are offered for sale as winter decoration. At the majority of these businesses and programs, funds that are received are allocated back into the program or utilized as salaries for the employees.

There are many opportunities to include these types of businesses in your region. Instead of buying pens, signs, brochures, or gifts from a corporate entity, look to see if there are social enterprises near you that can offer similar products. This way your organization can connect with a business or program that employs people with autism, establish a relationship with this type of business, which may want to join your Inclusion Team, and support them financially by making a purchase that is socially responsible and inclusive of your own customers. Below there are pictures of a candle from ScentsAbility Candles (www.scentsability.org) and a box of chocolate confections from The Chocolate Spectrum (www.thechocolatespectrum.com).

Agency Fair

A beneficial way to support your local disability programs, organizations, and service providers is to host an agency fair. This helps them share what they do with the community, and is an opportunity for your customers to learn about what resources are available for them. This type of event can be offered as a sole event where it is focused specifically on your local non-profits, or it can be done in association with one of your annual events or a specialized event that is designed for the disability community. For example, a job shadowing event is an excellent opportunity for young adults to explore different jobs and roles that are available at businesses. When combining this with an agency fair, it links employment services providers to individuals and can help them develop their pre-employment skills.

The agency fair aspect of this type of event would include setting out tables and tents for different agency representatives to share information about their programs, and to pass out their materials to your local customers. A business might provide a cloth bag or backpack for participants to use to hold all the items they collect, and for the business to include some of their materials and information on upcoming events. These types of programs can be extremely beneficial as it brings agencies that might be housed slightly further outside of your community, but that serve your customers, together, and that information goes directly to them. Agency fairs typically have minimal cost and can be both educational and fun for your business.

First Responder Event

Community events like Touch a Truck, where families come to meet first responders come in an informal and fun environment, can help decrease some of the stigma that may be associated with an encounter with emergency services and law enforcement personnel. It can also support a person with autism in desensitizing from some of the noises of fire trucks and police cars. A way to help make this type of event more disability friendly would be to follow some of the suggestions in the event planning portion of this book, but also to include opportunities for first responders to talk through and possibly even demonstrate some of what happens when they are called out to a scene. For example, a firefighter might show how they use the hose, or the police unit might bring out their search dogs to meet the public. For older children and teens, including some adults, it can be an opportunity to help them with some of the independent living skills related to emergencies that they might be slightly concerned about. For example, putting together a checklist of how to respond after a car accident if you were driving. This might be something to create and hand out with some guidance from law enforcement in this type of setting. Another option could be screening a safety movie together, such as Be Safe the Movie, which reviews appropriate ways to interact with law enforcement. Here is the link to the movie's website, and a short description from that site. https://besafethemovie.com

> *What would your son, daughter or student do in an encounter with the police? Run? Fight? Melt down? Those are scary possibilities. More news stories than ever focus attention on individuals with disabilities who have unsafe or even disastrous encounters with law enforcement. Now is the time to help young people with autism or other learning differences learn the skills they need to interact safely with the police. BE SAFE the Movie and BE SAFE Teaching Edition are the indispensable teaching tools you've been looking for! The BE SAFE video modelling program shows teens and adults what to do in everyday situations with the police. Actors with autism and other disabilities interact with real police officers, modelling safe words and actions. Whether it is being pulled over while driving or following instructions from the police, the 1-hour BE SAFE DVD teaches life-saving skills that everyone needs to know!*

Economic/Business Initiative

Connecting with other businesses can be a good way of deepening the commitment and broadening the scope of your Autism Friendly Business initiative. Inclusion programs that create partnerships with local companies are one way to accomplish this. These partnerships will require training and education for business owners and their employees. The businesses themselves can implement complementary Autism Friendly Business strategies, as you work together to align them with the inclusionary values of your specific business and the needs of the customers in your community.

Infrastructure is strengthened when you create inclusionary partnerships. The businesses you work with will support you as you support them. Develop a program and create some structure within that program in order to have to follow, strengthen the community bond through both alignments of goals and accountability. A business inclusion program in your region might include the following types of requirements to participate and maintain membership. Trainings commitments are needed, that include initial introduction to the program and more in-depth workshops about autism. There may be trainings throughout the year or when new employees are hired, which can be incorporated into onboarding and orientation. Other conditions might include hiring individuals with autism, or developing pre-employment support which can be in the form of mentoring, or having events such as job shadowing days, or possibly vocation training for people with autism that are interested in a specific career path. It might also include that the businesses participate in volunteering or contributing time to be on a task force or community board.

Your business will want to develop ways to acknowledge those partners so that they can be seen and accessed by the community. It is likely that people in the autism community are more frequently going to choose to go into an Autism Friendly

Business if given the opportunity. Businesses may also want to create a logo, certificate, or window cling that indicates that the business is part of the business's inclusion program.

Autism Awareness Month Proclamation and 'Light It Up Blue'

There are many ways of displaying autism awareness at your business. During World Autism Awareness Day or Autism Awareness Month in April, many places will 'light it up blue.' This means that they will either light their building, a sign, a pathway, or another prominent or symbolic structure, with blue lights to show their support for the autism community. This lighting often goes hand in hand with a ceremony or celebration to highlight the importance of this dedication. This is one very visible way to demonstrate your commitment to the neurodiverse customers in your region.

If you are connected to a government organization, cities have the benefit of creating proclamations, or specialized days and celebrations through directives from leadership. Sometimes Mayors or City Commissioners will even hand out a key to the city. These events are opportunities to recognize individuals with autism and the programs and non-profits that they may participate in or that support them. When celebrating autism awareness events, be sure to share your new initiatives and progress your business is making towards being Autism Friendly.

Other ways of increasing autism awareness are to boost the visibility of individuals with autism through your business practices. For example, when you share stories of participants or customers, or when you highlight customers in a newsletter or mailing, be sure to include a diverse representation of people, including those with autism. These could be success stories or perhaps times where your customers are engaged with your business. When you consider your approach, understand that visibility is important on many levels. The people being included will feel seen and recognized as a member of your community, other people with autism may feel a kinship or belonging when seeing your business being inclusive, and those without autism or that have a minimal understanding or knowledge of people with autism will learn more and be reminded that people with all sorts of gifts and differences live nearby.

Fundraise for Your Initiative and Support Others

There are some costs associated with developing and implementing Autism Friendly strategies. Costs can be minimal, but if you chose to create a marketing campaign to spread the word, or to remodel a specific space to incorporate a low-sensory room, there may be opportunities to fundraise to support those expenses. Businesses may also decide to commit to offer some financial support or fundraising for the organizations that support them in their consultative teams.

If your organization is planning to fundraise, either for your own programs, or to support your local disability non-profit or community agencies, these are a few ideas and suggestions on how to approach that objective. If you are going to fundraise for a non-profit, begin by asking what they need, and when they need it. For example, collecting or sponsoring back-to-school supplies and backpacks at the start of a new school year can be a supportive and helpful financial and outreach offering. A different option is to determine what month of the year is recognized as their celebration or tribute month. For example, in the United States, April is Autism Awareness Month (some groups are shifting the name to Autism Acceptance Month or Autism Celebration Month), and April 2nd is World Autism

Awareness Day. Other times of the year to celebrate with fundraising might be the annual anniversary of the Americans with Disabilities Act which takes place in July, or in October which is National Disability Employment month. Events can act as sponsorship and in-kind donations, which might include, in October specifically, a job fair where employers are committed to hire individuals with disabilities, job shadowing for young adults, or employment-related mentoring.

Fundraising for your initiative might be accomplished in ways that are more traditional, such as a gala with a silent auction, or a 5K run. Other ways of raising money that also help to raise a company-wide awareness of new initiatives could be to offer staff certain opportunities to donate where they get to take part in a fun activity or a treat, within a lens of the new program. Some businesses have had a group creativity workshop where the proceeds are donated, or they've had dress down day where staff get to wear casual attire to work for their donation. Some businesses even offer community-based raffles. It may require some technical support internally, but there are organizations that have opportunities for employees to engage with charitable donations that come out of directly a pay check, which can be something that is advertised and set up annually. Other groups may offer an opportunity for customers to donate either by having boxes placed around their lobby or cash register, or in some cases if individuals are paying to participate in a program, a recreational activity, or even a meal at a restaurant, they can round up to the nearest dollar, and the extra amount that they pay becomes a donation.

In whichever way you choose to go about fundraising for your own programs, or to support your local organizations and agencies, know that it will be appreciated and is a way to deepen your connection with those that contribute to your initiatives, as well as your customers, sharing how much their time and effort has meaning and that inclusion is an area your city is dedicated to and that it is meaningful.

Create a Website or Social Media Page

The only way your community will know that your business is working on Autism Friendly and inclusion initiatives is if you tell them. One of the best ways to communicate with your customers is through digital platforms. People come together on social media, utilize apps for navigation and information gathering, and seek solutions from your website. An excellent way to highlight what products, services, and opportunities you offer is to create an inclusion-specific digital media campaign. You may decide to create a unique Autism Friendly or inclusion-specific logo as a visual. Some businesses work with their marketing departments and develop materials, program names or titles, and press releases. After creating an exclusive way of showcasing your initiatives, share it across platforms and communication materials for your customers to understand your commitment to inclusion, and for them to plug into your programs and engage. Also, be sure to work with your inclusion team and ask them to share throughout their own networks. Keeping these platforms updated with new initiatives, events, and general autism-related information can be useful and informative, with some businesses even creating a newsletter for this purpose.

Below, you will find website and social media links, as examples of marketing Autism Friendly businesses, cities, and programs. As you can see they vary in presentation of how they market and share information about being Autism Friendly. The cities are the City of Boynton Beach, the City of Tampa, and the City of Miami. The businesses are Royal Caribbean Cruise Line and Zoo Tampa. Using platforms that directly communicate to your customers is an efficient and effective way to share information, forms, social stories, and visual supports with your customers and potential customers. It helps them understand what you are able to offer and what information they will need to provide to have a supportive experience with your company. Sharing your initiative also creates online resources, models, and samples for businesses at the beginning of the process. This fosters growth and supports broadening this movement. It is possible that as you bring your efforts to the attention of the community, that it may inspire and motivate a business on the other side of the world to commit to inclusion and begin the process of increasing accessibility to the neurodiverse community.

https://www.facebook.com/InclusiveBoynton/
https://www.tampa.gov/accessibility/autism-friendly-tampa
https://accessmiami.org/
https://www.royalcaribbean.com/experience/accessible-cruising/autism-friendly-ships
https://zootampa.org/visitor-info/accessibility/

Index

For Product Safety Concerns and Information please contact our EU representative GPSR@taylorandfrancis.com
Taylor & Francis Verlag GmbH, Kaufingerstraße 24, 80331 München, Germany

www.ingramcontent.com/pod-product-compliance
Ingram Content Group UK Ltd.
Pitfield, Milton Keynes, MK11 3LW, UK
UKHW022120190926
475702UK00013B/1561

* 9 7 8 1 0 3 2 0 2 8 3 1 6 *